MW00889626

NO GRID SURVIVAL PROJECTS BIBLE

Be Ready to Thrive through Crisis, Disasters, and Blackouts -The Ultimate DIY Guide for a Fortified Home, Dependable Power, and Plentiful Food Supply [11 books in 1]

Robert Harrison

In loving memory of my father,
Your guidance led my first steps into nature's vast embrace. This book journeys back,
Celebrating the survival spirit you ignited within me. Your spirit guides me still.

Selected Array of Bonus Videos for Readers!

Dive deep into the essence of self-reliance with my book, "No Grid Survival Projects Bible," and unlock a treasure trove of knowledge that stands ready to guide you into a life of independence. Your commitment to preparing for the unexpected and living a self-sufficient life deserves recognition and support. That's why I've prepared an exclusive gift just for you: **31 engaging bonus videos that perfectly complement the enriching lessons of the book.**

These videos are more than just visual aids; they are a carefully selected collection designed to bring the book's teachings to life right before your eyes. From the convenience of your device, you'll see theories transform into tangible skills. Each video, optimized for clarity on any screen, offers detailed, step-by-step instructions on key projects that will significantly enhance your experience of living off the grid.

Unlocking this treasure trove of knowledge is effortless—simply scan the QR code provided with your smartphone or tablet. Optimized for convenient viewing on any device, these resources allow you to dive deeper into self-sufficiency at your own pace, wherever you are. This integrated approach, combining the depth of the written word with dynamic video demonstrations, promises a comprehensive and engaging guide to mastering the art of living independently.

So, grab your device, scan the QR code, and embark on this interactive voyage of discovery and empowerment. Welcome aboard your enhanced journey to mastering self-sufficiency, where every page turned and every video watched brings you closer to a life of independence and resilience. Let's animate these pages together and transform knowledge into action!

BECOME SELF-SUFFICIENT. DON'T JUST SURVIVE, THRIVE!

SCAN ME

The material contained in this publication is protected by copyright law and may not be reproduced, duplicated, or transmitted in any form or by any means without the explicit written permission of the author or publisher. The author and publisher shall not be held liable for any damages, reparations, or financial losses resulting from the use of the information provided in this publication, whether directly or indirectly.

Legal Notice

This publication is safeguarded by copyright laws and is intended for personal use only. Any unauthorized modification, distribution, sale utilization, citation, or rephrasing of any part or content from this publication is strictly prohibited without the express consent of the author or publisher.

Disclaimer

Please note that the information contained in this publication is for educational and entertainment purposes only. While every effort has been made to ensure the accuracy, timeliness, and reliability of the information presented, no warranties of any kind, whether explicit or implied, are provided. Readers are advised that the author does not offer legal, financial, medical, or professional advice. The content of this publication is sourced from various references, and readers should seek advice from a qualified professional before applying any techniques or practices outlined herein. By accessing this publication, the reader agrees that the author shall not be held liable for any direct or indirect losses resulting from the use of the information contained herein, including but not limited to errors, omissions, or inaccuracies.

Copyright © 2024 By Robert Harrison
NO GRID SURVIVAL PROJECTS BIBLE
All Right Reserved

Table of Contents

INTRODUCTION..14

BOOK 1: PREPPING AND SURVIVALISM GUIDE.................................17

CHAPTER 1: INTRODUCTION TO SURVIVALISM............................18

CHAPTER 2: PREPPING, BUSHCRAFTING, AND SURVIVALISM 20

CHAPTER 3: TEOTWAWKI AND SHTF SCENARIOS.........................23

CHAPTER 4: WORST-CASE SCENARIOS 26

    NATURAL DISASTERS 26

    PANDEMICS 27

    ECONOMIC COLLAPSE 27

    POWER OUTAGES 28

    CIVIL UNREST 28

    TECHNOLOGICAL BREAKDOWN 28

    CLIMATE CHANGE IMPACTS..29

    NUCLEAR ACCIDENTS 30

    CYBERSECURITY THREATS 30

    GEOPOLITICAL TENSIONS..30

CHAPTER 5: BUGGING IN VS. BUGGING OUT 32

CHAPTER 6: WATER, FIRE, AND FOOD IN SURVIVAL 35

    WATER 35

    FIRE..36

    FOOD 37

    INTEGRATION OF WATER, FIRE, AND FOOD IN SURVIVAL.......38

CHAPTER 7: CANNING AND FOOD PRESERVATION.......................40

    STEP-BY-STEP GUIDE:...40

    TIPS FOR SUCCESSFUL CANNING:..43

CHAPTER 8: MENTAL HEALTH IN CRISES 45

UNDERSTANDING THE IMPACT OF CRISES ON MENTAL HEALTH.................................45

BUILDING RESILIENCE 47

CHAPTER 9: SHELTER ESSENTIALS 49

TYPES OF SHELTERS..49

    Tents 49

    Tarps and Hammocks 50

    Emergency Shelters 50

    Natural Shelters 50

    Snow Shelters 50

SHELTER CONSTRUCTION BASICS..50

    Site Selection 50

    Materials..50

    Anchoring and Stability 51

    Insulation...51

    Ventilation 51

SITUATIONAL APPROPRIATENESS..51

    Camping Trips..51

    Backpacking 51

    Survival Scenarios 51

    Winter Survival 51

    Emergency Situations 51

    Outdoor Exploration 51

BOOK 2: WORST CASE SCENARIO PRACTICAL GUIDE 52

CHAPTER 1: **SURVIVING A NUCLEAR DISASTER**..53

CHAPTER 2: **NAVIGATING A PANDEMIC CRISIS** 55

CHAPTER 3: **SURVIVING NATURAL DISASTERS**...57

CHAPTER 4: **SURVIVING POWER GRID FAILURES** 59

CHAPTER 5: COPING WITH URBAN DISRUPTIONS 60

BOOK 3: HOME DEFENSE GUIDE 61

CHAPTER 1: SECURING THE HOME 62

CHAPTER 2: PERIMETER DEFENSE 65

CHAPTER 3: HOME HARDENING 69

CHAPTER 4: SECRET STORAGE SOLUTIONS 73

CHAPTER 5: PROTECTING VULNERABLE FAMILY MEMBERS 76

    CHILD SAFETY 76

    ELDERLY SAFETY 76

    PERSONALIZED SAFETY MEASURES 77

    FIRE SAFETY MEASURES 78

    WEATHER AND NATURAL DISASTER PREPAREDNESS 78

CHAPTER 6: ROLE OF GUARD DOGS 80

CHAPTER 7: COMMUNICATION PROTOCOLS 84

CHAPTER 8: OUTSIDE EXCURSIONS 88

BOOK 4: SURVIVAL SHELTER GUIDE 92

CHAPTER 1: SHELTER CONSTRUCTION 93

    LEAN-TO SHELTER 93

    DEBRIS HUT SHELTER 93

    A-FRAME SHELTER 94

    SNOW CAVE SHELTER 95

    TARP SHELTER 95

    TEEPEE SHELTER 96

    BIVY BAG SHELTER 96

CHAPTER 2: SHELTER HARDENING 97

CHAPTER 3: WILD SHELTERS 100

CHAPTER 4: ANTI-ATOMIC SHELTERS 105

    KEY COMPONENTS OF ANTI-ATOMIC SHELTERS 105

    BUILDING AN ANTI-ATOMIC SHELTER 106

    MAINTAINING AND UPDATING THE BUNKER 107

CHAPTER 5: DIY BUNKERS 108

BOOK 5: SELF-DEFENSE AND FIRST AID CRASH COURSE 112

CHAPTER 1: **FUNDAMENTALS OF SELF-DEFENSE** 113

CHAPTER 2: **CREATING A SURVIVAL KIT** 116

    SHELTER AND WARMTH 116

    WATER AND HYDRATION 116

    FOOD AND NUTRITION 116

    NAVIGATION AND COMMUNICATION 117

    FIRST AID AND MEDICAL SUPPLIES 117

    TOOLS AND MULTI-FUNCTION ITEMS 117

    ILLUMINATION 117

    CLOTHING AND PROTECTION 118

    PERSONAL HYGIENE ITEMS 118

    PERSONAL DOCUMENTS AND IDENTIFICATION 118

CHAPTER 3: **FIRST AID BASICS** 119

CHAPTER 4: **ESSENTIAL MEDICINE INVENTORY** 123

    PAIN RELIEF AND FEVER REDUCERS 123

    ANTIHISTAMINES 124

    GASTROINTESTINAL MEDICATIONS 124

    ADHESIVE BANDAGES AND ANTISEPTIC OINTMENT 124

    THERMOMETER 124

    COUGH AND COLD MEDICATIONS 125

    ORAL REHYDRATION SOLUTIONS 125

    ALLERGY MEDICATIONS 125

Topical Corticosteroids 125

Oral Pain Relief 125

Eye Drops 126

Motion Sickness Medications 126

Anti-fungal Cream 126

Prescription Medications 126

Personalized Medications 127

Storage and Safety Tips: 127

CHAPTER 5: **MENTAL HEALTH MANAGEMENT** **128**

BOOK 6: THE PERFECT PREPPER'S SURVIVAL CHECKLIST **130**

CHAPTER 6: **SURVIVAL CHECKLIST** **131**

Water Supply 131

Food and Nutrition 131

Shelter and Warmth 131

First Aid and Medical Supplies 132

Tools and Multi-Function Items 132

Communication Devices 132

Personal Hygiene Items 132

Clothing and Protection 133

Financial and Legal Documents 133

Self-Defense and Security 133

Navigation and Maps 133

Entertainment and Comfort 133

Extra Supplies for Children and Pets 134

Maintenance and Repair Items 134

Mental Health Support 134

BOOK 7: WATER MASTERY **135**

CHAPTER 1: **WATER SOURCING TECHNIQUES** **136**

CHAPTER 2: DIY PURIFICATION SYSTEMS **139**

CHAPTER 3: WATER STORAGE SOLUTIONS **142**

LONG-TERM WATER STORAGE SOLUTIONS 142

WATER CONSERVATION STRATEGIES 143

BOOK 8: SUSTENANCE AND NUTRITION **145**

CHAPTER 1: FORAGING GUIDE **146**

EDIBLE PLANTS 146

    Dandelion (Taraxacum officinale) 146

    Wild Garlic (Allium ursinum) 147

    Stinging Nettle (Urtica dioica) 147

    Chickweed (Stellaria media) 147

    Plantain (Plantago major) 147

    Wood Sorrel (Oxalis spp.) 147

    Burdock (Arctium lappa) 147

    Wild Strawberries (Fragaria vesca) 147

    Cattail (Typha spp.) 147

    Mallow (Malva spp.) 147

    Acorns (Quercus spp.) 148

    Pine Needle Tea (Pinus spp.) 148

EDIBLE INSECTS 148

    Crickets 148

    Mealworms (Darkling Beetle Larvae) 148

    Grasshoppers 148

    Ants 148

    Termites 148

    Silkworm Pupae 149

    Beetle Larvae (e.g., Rhinoceros Beetle Grubs) 149

    Butterfly and Moth Larvae (Caterpillars) 149

    Dragonfly and Damselfly Nymphs 149

Honeybee Larvae *149*

SAFETY CONSIDERATIONS 149

CHAPTER 2: **ETHICAL HUNTING PRACTICES** **150**

UNDERSTANDING ETHICAL HUNTING 150

SUSTAINABLE PRACTICES 151

RESPECTFUL AND HUMANE PRACTICES 151

PERSONAL ETHICS AND RESPONSIBILITY 152

CHAPTER 3: **FOOD PRESERVATION METHODS** **153**

FREEZE DRYING 153

FERMENTATION 153

CANNING WITH PRESSURE 153

CONTROLLED ATMOSPHERE STORAGE 154

DEHYDRATION WITH TECHNOLOGY 154

HIGH-PRESSURE PROCESSING (HPP) 154

VACUUM PACKING 154

RADIATION PRESERVATION 155

OSMOTIC DEHYDRATION 155

BOOK 9: FIRE AND COOKING **157**

CHAPTER 1: **FIRE CREATION TECHNIQUES** **158**

PRIMITIVE FIRE-STARTING TECHNIQUES 158

FLINT AND STEEL 159

FIRE PISTONS 159

MAGNIFYING GLASS OR LENS 160

FIRE STARTERS AND FERROCERIUM RODS 160

BATTERY AND STEEL WOOL 160

FUELING AND MAINTAINING THE FIRE 161

CHAPTER 2: **OFF-GRID COOKING SOLUTIONS** **162**

OUTDOOR ROCKET STOVE 162

SOLAR COOKER 163

DIY PORTABLE CAMPING STOVE 163

OUTDOOR CLAY OVEN (COB OVEN) 163

DIY CAMPFIRE GRILL 164

PROPANE CAMPING STOVE 164

HAYBOX COOKER 164

DIY WOOD-FIRED PIZZA OVEN 165

CHAPTER 3: **SURVIVAL RECIPES** **166**

FORAGED VEGETABLE STIR-FRY 166

CAMPFIRE GRILLED FISH 166

NETTLE AND ACORN SOUP 167

WILD TEA INFUSION 168

SAUTÉED CATTAIL SHOOTS 168

ROASTED ROOT VEGETABLES 169

HICKORY NUT TRAIL MIX 169

WILDFLOWER SALAD 170

WILD MUSHROOM RISOTTO 170

ACORN PANCAKES 171

CATTAIL POLLEN PANCAKES 171

DRIED FRUIT LEATHER 172

WILD MINT TEA AND BISCUITS 172

ROSEHIP AND PINE NEEDLE INFUSED TEA 173

SASSAFRAS ROOT BEER 173

WILD EDIBLE FLOWER FRITTERS 174

SPRUCE TIP SYRUP 175

AMARANTH AND PURSLANE SALAD 175

HAZELNUT AND MAPLE ENERGY BARS 176

BIRCH SAP LEMON SORBET 176

BOOK 10: NAVIGATION AND COMMUNICATION **178**

CHAPTER 1: ADVANCED NAVIGATION SKILLS 179

CHAPTER 2: EMERGENCY SIGNAL CRAFTING 182

VISUAL SIGNALS 182

AUDITORY SIGNALS 182

COMMUNICATION THROUGH OBJECTS 183

TECHNOLOGY-ASSISTED SIGNALS 183

ENVIRONMENTAL ADAPTATION 184

PERSISTENCE AND CONSISTENCY 184

CHAPTER 3: TECHNOLOGY IN SURVIVAL 185

GPS TECHNOLOGY IN THE WILDERNESS 185

RADIOS AS LIFELINES IN THE WILDERNESS 186

CHALLENGES AND CONSIDERATIONS 187

INTEGRATING TECHNOLOGY INTO YOUR SURVIVAL PLAN: 187

BOOK 11: TOWARDS COMPLETE SELF-SUFFICIENCY 189

CHAPTER 1: TRANSITIONING TO OFF-GRID LIVING 190

CHAPTER 2: RENEWABLE ENERGY PROJECTS 193

SOLAR POWER SYSTEM 193

WIND TURBINE SYSTEM 194

HYDROELECTRIC POWER SYSTEM 194

CHAPTER 3: DESIGNING A SUSTAINABLE HOMESTEAD 196

CONCLUSION 199

Introduction

The purpose of this book is to provide a comprehensive guide on prepping, survivalism, and self-sufficiency in the face of various worst-case scenarios. This book covers a wide range of topics related to survival, including prepping basics, shelter construction, worst-case scenario practical guides, home defense, survival shelters, self-defense, first aid, water mastery, sustenance and nutrition, fire and cooking, navigation and communication, and transitioning to off-grid living with renewable energy projects.

- **Book 1: Prepping and Survivalism Guide** - This section introduces the reader to the concepts of survivalism and prepping. It covers various scenarios, including TEOTWAWKI (The End of the World as We Know It) and SHTF (Sh*t Hits the Fan) situations. It explores worst-case scenarios such as natural disasters, pandemics, economic collapse, power outages, civil unrest, and more.
- **Book 2: Worst Case Scenario Practical Guide** - This part delves into practical survival strategies for specific worst-case scenarios, including nuclear disasters, pandemics, natural disasters, power grid failures, and coping with urban disruptions.
- **Book 3: Home Defense Guide** - Focused on securing and defending one's home, this section covers topics like securing the home, perimeter defense, home hardening, secret storage solutions, and protecting vulnerable family members.
- **Book 4: Survival Shelter Guide** - Provides detailed information on shelter construction, hardening, and different types of shelters, including wild shelters and anti-atomic shelters. It also covers DIY bunkers.
- **Book 5: Self-Defense and First Aid Crash Course** - Covers the fundamentals of self-defense, creating a survival kit, first aid basics, essential medicine inventory, and mental health management.
- **Book 6: The Perfect Prepper's Survival Checklist** - Presents a comprehensive checklist for survival, covering water supply, food and nutrition, shelter and warmth, first aid and medical supplies, tools, communication devices, personal hygiene items, clothing and protection, legal documents, self-defense and security, navigation, and more.
- **Book 7: Water Mastery** - Focuses on water sourcing techniques, DIY purification systems, and water storage solutions.
- **Book 8: Sustenance and Nutrition** - Covers foraging guides, ethical hunting practices, food preservation methods, and edible plants and insects.

- **Book 9: Fire and Cooking** - Explores fire creation techniques, off-grid cooking solutions, and survival recipes.
- **Book 10: Navigation and Communication** - Provides advanced navigation skills, emergency signal crafting, and discusses the role of technology in survival.
- **Book 11: Towards Complete Self-Sufficiency** - Addresses transitioning to off-grid living, renewable energy projects, and designing a sustainable homestead.

This guide caters to individuals intrigued by the concepts of prepping and survivalism across diverse scenarios. It delves into an extensive array of subjects associated with survival, encompassing preparation for natural disasters, pandemics, economic downturns, power failures, civil unrest, technological disruptions, climate change consequences, nuclear incidents, cybersecurity risks, and geopolitical tensions.

Tailored for those interested in or concerned about readiness for various emergencies, disasters, or societal upheavals, the book is designed for individuals seeking to acquire the knowledge and skills necessary to navigate challenging situations. It aims to assist in securing homes and achieving a level of self-sufficiency concerning essential resources such as water, food, and energy. Whether you are a novice seeking an introduction to survivalism or an experienced prepper seeking detailed information on specific aspects of preparedness, this book provides valuable insights.

Preparedness is crucial for ensuring individuals and communities can withstand and overcome challenges in the absence of traditional power sources. In such situations, where the electrical grid fails, people must be self-reliant to meet their basic needs.

- **Energy Independence:** Preparedness ensures alternative energy sources like solar panels, generators, or wind turbines, reducing dependence on the grid.
- **Water Security:** Storing water and having purification methods in place guarantees a stable water supply during grid failures or disasters.
- **Food Resilience:** Stockpiling non-perishable food items and cultivating a home garden ensures a sustainable food source when traditional supply chains are disrupted.
- **Communication Strategies:** Establishing communication plans using radios, satellite phones, or other non-grid-dependent methods is crucial for staying informed and connected during emergencies.
- **Medical Supplies:** A well-equipped medical kit and knowledge of basic first aid become essential in the absence of immediate access to hospitals or pharmacies.
- **Self-Defense Measures:** In unpredictable situations, having knowledge of self-defense and possessing necessary tools ensures personal safety.

- **Community Collaboration:** Building a network with neighbors and local communities fosters mutual support during crises, creating a more resilient environment.
- **Skill Development:** Acquiring survival skills such as fire-making, navigation, and basic carpentry enhances adaptability and self-sufficiency.
- **Financial Preparedness:** Keeping cash reserves and valuable assets secure allows for continued transactions when electronic financial systems fail.
- **Shelter Planning:** Establishing a secure and sustainable shelter, whether at home or in a designated location, is vital for protection against the elements.
- **Continuous Education:** Staying informed about potential threats, learning new skills, and adapting to evolving situations ensures ongoing readiness for unforeseen challenges.

Preparedness involves a holistic approach encompassing energy, water, food, communication, medical, defense, community, skills, finance, and shelter aspects, ensuring individuals can navigate and thrive in the absence of a functioning grid.

BOOK 1:
PREPPING AND
SURVIVALISM GUIDE

Chapter 1:

Introduction to Survivalism

Survivalism is more than just a set of skills or practices—it's a mindset that revolves around preparedness, self-sufficiency, and adaptability in the face of unexpected challenges. In a world that often seems unpredictable and dynamic, the concept of survivalism has gained traction as individuals seek ways to navigate uncertain times.

Survivalism is a philosophy and way of life rooted in the belief that individuals should be prepared for unforeseen circumstances, ranging from natural disasters to societal disruptions. At its essence, survivalism is not about extreme pessimism but rather a realistic acknowledgment that unexpected events can occur, and being prepared can mitigate their impact. It involves acquiring a diverse set of skills, knowledge, and resources to thrive in various situations.

Survivalism can manifest in different forms, from basic preparedness for power outages to more elaborate plans for extended periods without external support. The goal is to be self-reliant and self-sufficient, reducing dependence on external systems and resources.

The Survivalist Mindset

Central to survivalism is the development of a resilient mindset—one that embraces adaptability, resourcefulness, and a proactive approach to challenges. This mindset involves recognizing that change is inevitable and being prepared to respond effectively. Here are key components of the survivalist mindset:

1. **Adaptability:** Survivalists understand that circumstances can change rapidly. Being adaptable means being open to new information, flexible in problem-solving, and quick to adjust plans as needed.

2. **Resourcefulness:** A survivalist mindset values resourcefulness—making the most of available resources and finding creative solutions to problems. It involves repurposing items, improvising tools, and maximizing efficiency.

3. **Preparedness:** Preparedness is the cornerstone of survivalism. This involves anticipating potential challenges and taking proactive measures to mitigate risks. It includes acquiring essential skills, building a stockpile of supplies, and having contingency plans in place.

4. **Self-Reliance:** Survivalists prioritize self-reliance, aiming to reduce dependence on external systems. This involves learning practical skills such as basic first aid, foraging for food, purifying water, and mastering survival techniques.

5. **Community Building:** While self-reliance is crucial, survivalism doesn't exclude the importance of community. Building a network of like-minded individuals fosters mutual support and strengthens collective resilience.

Key Principles of Survivalism

1. **Assessment of Risks:** Survivalism begins with a realistic assessment of potential risks. This involves considering geographical location, climate, and the likelihood of specific events like natural disasters or social unrest.

2. **Skill Development:** Acquiring a diverse set of skills is fundamental to survivalism. This includes learning first aid, navigation, shelter building, food preservation, and other practical abilities.

3. **Resource Stockpiling:** Survivalists often build reserves of essential supplies, including non-perishable food, water, medical supplies, and tools. These stockpiles act as a buffer during times of scarcity or disruption.

4. **Continuous Learning:** The survivalist mindset encourages a commitment to continuous learning. Staying informed about new technologies, survival techniques, and developments in various fields ensures adaptability in evolving situations.

Chapter 2:

Prepping, Bushcrafting, and Survivalism

In a world where uncertainties abound, individuals often turn to various practices to enhance their preparedness and self-sufficiency. Three such endeavors—prepping, bushcrafting, and survivalism—though distinct in their focus, share a common thread of self-reliance and adaptability.

Prepping

Prepping, short for preparedness, is a lifestyle centered around anticipating and readying oneself for a range of potential disruptions or disasters. Preppers, individuals who adhere to this philosophy, believe in proactive planning to mitigate the impact of unforeseen events. The focus extends beyond survival scenarios, encompassing various aspects of life, from natural disasters to economic downturns.

Key elements of prepping include:

1. **Stockpiling Resources:** Preppers often accumulate supplies like non-perishable food, water, medical provisions, and other essentials to ensure a sufficient buffer during times of scarcity.

2. **Emergency Planning:** Prepping involves creating comprehensive plans for various emergencies. This includes evacuation strategies, communication plans, and knowing how to access or create safe havens.

3. **Skill Acquisition:** Preppers prioritize acquiring a diverse set of skills, such as first aid, navigation, and basic survival techniques, to enhance their ability to adapt to different scenarios.

4. **Financial Preparedness:** Beyond physical resources, prepping may involve financial preparations, such as savings and investments, to withstand economic uncertainties.

Bushcrafting

Bushcrafting focuses on developing primitive skills and traditional craftsmanship to thrive in natural environments. It's a hands-on approach to living in the wilderness, emphasizing the use of natural materials and self-sufficiency.

Key components of bushcrafting include:

1. **Primitive Skills:** Bushcrafting involves mastering traditional skills like fire making, shelter building, and foraging for food. This knowledge is often derived from indigenous practices and historical survival techniques.

2. **Connection with Nature:** Bushcrafters seek a profound connection with the natural world. They learn to read the environment, identify edible plants, and use the resources provided by nature to meet their needs.

3. **Tool Crafting:** Bushcrafting places a strong emphasis on crafting tools from natural materials. This includes making knives, bows, and other implements using techniques passed down through generations.

4. **Low-Impact Living:** Bushcrafters often adopt a low-impact lifestyle, minimizing their environmental footprint and respecting the balance of nature.

Survivalism

Survivalism, as discussed in the previous section, is a mindset and way of life rooted in being prepared for unexpected challenges. Survivalists focus on adaptability, resourcefulness, and self-reliance, preparing for a wide range of scenarios that may require immediate action.

Key principles of survivalism include:

1. **Assessment of Risks:** Survivalists assess potential risks, considering factors like geographic location, climate, and societal conditions to prepare for various contingencies.

2. **Skill Development:** Acquiring a diverse set of practical skills is fundamental to survivalism, covering areas such as first aid, navigation, and shelter building.

3. **Resource Stockpiling:** Similar to prepping, survivalism often involves accumulating reserves of essential supplies, ensuring a degree of self-sufficiency during challenging times.

4. **Continuous Learning:** The survivalist mindset promotes a commitment to ongoing learning, staying informed about new technologies, survival techniques, and developments in various fields.

Synergies and Overlaps

While prepping, bushcrafting, and survivalism each have distinct focuses, there are noteworthy areas where these pursuits intersect, creating synergies that can enhance an individual's overall preparedness and self-sufficiency:

1. **Skill Transfer:** The skills acquired in bushcrafting, such as fire-making and shelter-building, can be directly applicable in survivalist scenarios. Likewise, the practical skills learned in prepping and survivalism can complement the hands-on abilities of a bushcrafter.

2. **Resource Utilization:** Bushcrafting emphasizes using natural resources efficiently, a principle that aligns with both prepping and survivalism. Learning to make the most of available materials and resources enhances self-sufficiency in all three domains.

3. **Environmental Awareness:** All three pursuits foster a heightened awareness of the environment. Whether it's understanding the risks associated with a specific location, identifying edible plants, or respecting the balance of nature, environmental consciousness is a common thread.

4. **Self-Reliance:** At the core of prepping, bushcrafting, and survivalism is the value of self-reliance. While the specific approaches may differ, the shared goal is to reduce dependence on external systems and enhance personal resilience.

Chapter 3:

TEOTWAWKI and SHTF Scenarios

Within preparedness and survivalism, two acronyms frequently emerge with a sense of seriousness: TEOTWAWKI (The End of The World as We Know It) and SHTF (S**t Hits the Fan). These expressions capture the expectation of extraordinary occurrences that might disturb established societal patterns and test our adaptability.

TEOTWAWKI

TEOTWAWKI encapsulates the notion of a scenario where the world undergoes transformative, irreversible changes. It suggests a paradigm shift in society, making old ways no longer relevant. While the term can evoke images of apocalyptic scenarios, it's essential to recognize that TEOTWAWKI is a spectrum encompassing various levels of disruption, from regional upheavals to global cataclysms.

Understanding TEOTWAWKI involves contemplating scenarios such as:

1. **Natural Disasters:** Catastrophic events like mega-earthquakes, tsunamis, or supervolcanic eruptions can lead to widespread destruction, impacting entire regions and challenging the resilience of affected communities.

2. **Pandemics:** Severe global pandemics, with high mortality rates and societal breakdown, fall under the TEOTWAWKI umbrella. The COVID-19 pandemic demonstrated the far-reaching consequences of a highly contagious and deadly virus.

3. **Economic Collapse:** A collapse of financial systems leading to hyperinflation, economic depression, and widespread unemployment could result in a societal shift, disrupting established structures and norms.

4. **Technological Breakdown:** A large-scale failure of critical infrastructure, such as power grids or communication networks, could plunge societies into chaos, creating a TEOTWAWKI scenario.

5. **Social and Political Unrest:** If tensions escalate to the point of widespread civil unrest, governments collapsing, and the breakdown of law and order, it could be considered a TEOTWAWKI event.

Preparing for TEOTWAWKI involves comprehensive planning, including:

1. **Long-Term Sustainability:** Building self-sufficiency through food production, water harvesting, and renewable energy sources to sustain oneself in a world where conventional systems might be non-functional.

2. **Community Building:** Establishing strong connections with like-minded individuals and communities to foster mutual support and resource-sharing during challenging times.

3. **Skill Acquisition:** Developing a diverse set of skills, ranging from basic survival techniques to more specialized abilities like medical training and engineering, to enhance adaptability in a transformed world.

4. **Strategic Location:** Choosing a location that minimizes exposure to potential risks, such as natural disasters or geopolitical tensions, and allows for sustainable living.

SHTF

SHTF scenarios refer to immediate, often unexpected situations where normalcy abruptly disintegrates. Unlike TEOTWAWKI, SHTF events might be more localized or short-term but can still be intense and disruptive. The term implies a state of emergency that demands rapid adaptation and resourcefulness.

Examples of SHTF scenarios include:

1. **Power Outages:** Widespread and prolonged power failures, whether due to natural disasters or intentional attacks on infrastructure, can lead to chaos, especially in urban areas.

2. **Civil Unrest:** Social or political upheaval, protests, or rioting can quickly escalate into SHTF situations, challenging personal safety and security.

3. **Supply Chain Disruptions:** Events like economic crises, trade disruptions, or pandemics can lead to shortages of essential goods, causing panic and potential breakdowns in social order.

4. **Natural Disasters:** While some natural disasters fall under TEOTWAWKI, others, such as hurricanes, tornadoes, or floods, can create immediate SHTF situations, demanding swift responses to ensure personal safety.

Preparedness for SHTF scenarios involves:

1. **Emergency Kits:** Possessing emergency kits that are easily available and include vital supplies such as water, food that does not perish, first aid equipment, and significant documents.

2. **Communication Plans:** Establishing effective communication channels with family members, neighbors, and community networks to share information and coordinate responses during emergencies.

3. **Security Measures:** Taking measures to secure one's home and possessions, including fortifying entry points, having backup power sources, and considering self-defense strategies.

4. **Evacuation Plans:** Being prepared to evacuate if necessary, with pre-established routes and designated meeting points for family members.

Synergies Between TEOTWAWKI and SHTF

While TEOTWAWKI and SHTF scenarios differ in scale and duration, they share commonalities that warrant a holistic approach to preparedness:

1. **Skill Overlap:** Many of the skills necessary for surviving TEOTWAWKI events, such as navigation, first aid, and self-defense, are also crucial in SHTF scenarios.

2. **Resource Management:** Whether facing a temporary disruption or a long-term societal shift, effective resource management is key. Skills learned in SHTF scenarios can be applicable in a TEOTWAWKI context.

3. **Community Resilience:** Building a resilient community is beneficial in both scenarios. The support networks established for SHTF events can evolve to provide long-term assistance in TEOTWAWKI situations.

4. **Adaptability:** The ability to adapt quickly is a common theme. Whether it's a sudden crisis or a prolonged societal shift, those who can adjust their strategies and plans are more likely to thrive.

Chapter 4:

Worst-Case Scenarios

In a world fraught with uncertainties, preparing for worst-case scenarios becomes a crucial aspect of personal and community resilience. From natural disasters to societal upheavals, various emergencies can disrupt normalcy and challenge our ability to adapt.

Natural Disasters

Natural disasters encompass a wide range of events such as earthquakes, hurricanes, floods, wildfires, tornadoes, and tsunamis. These events often result from natural forces and can cause widespread destruction.

Strategies:

1. **Early Warning Systems:** Stay informed through weather alerts and early warning systems to have advance notice of potential disasters.

2. **Emergency Kits:** Make sure that you have emergency kits ready with all of the necessary materials, including water, food that does not go bad, first aid supplies, and critical documents.

3. **Evacuation Plans:** Establish evacuation plans with predetermined routes and meeting points for family members.

Pandemics

A pandemic is the global outbreak of a contagious disease, often characterized by high transmission rates and significant societal impact. Examples include the H1N1 influenza pandemic and the more recent COVID-19 pandemic.

Strategies:

1. **Hygiene Practices:** Emphasize good hygiene practices, including regular handwashing and sanitation.

2. **Social Distancing:** Follow guidelines for social distancing to reduce the spread of contagious diseases.

3. **Stockpiling Essentials:** Maintain a stockpile of essential supplies, including medications and hygiene products.

Economic Collapse

When a nation's financial system experiences a catastrophic and unexpected failure, this is referred to as a financial crisis, leading to widespread unemployment, hyperinflation, and economic depression.

Strategies:

1. **Financial Preparedness:** Diversify assets and investments to withstand economic uncertainties.

2. **Skill Development:** Acquire diverse skills that can be valuable in different job markets.

3. **Emergency Fund:** Maintain an emergency fund to cover essential expenses during periods of financial instability.

Power Outages

Power outages occur when there is a disruption in the electrical grid, leaving homes and businesses without electricity. These can result from natural disasters, equipment failures, or intentional attacks.

Strategies:

1. **Alternative Power Sources:** Invest in alternative power sources such as generators, solar panels, or battery backups.

2. **Energy Conservation:** Implement energy-saving practices to prolong available power resources.

3. **Emergency Lighting:** Have emergency lighting solutions such as flashlights, candles, or lanterns readily available.

Civil Unrest

Civil unrest involves public protests, demonstrations, or rioting that escalate into widespread social and political chaos, challenging law and order.

Strategies:

1. **Security Measures:** Implement security measures to protect your home and possessions.

2. **Communication Plans:** Establish effective communication channels with family members and community networks.

3. **Evacuation Plans:** Have evacuation plans in place in case the situation becomes untenable.

Technological Breakdown

A technological breakdown occurs when critical infrastructure, such as power grids or communication networks, fails on a large scale, leading to disruptions in various aspects of daily life.

Strategies:

1. **Information Security:** Protect personal information through cybersecurity measures.

2. **Backup Systems:** Create backup systems for essential electronic devices and data.

3. **Skill Development:** Acquire skills that do not rely heavily on technology, enhancing adaptability in case of breakdowns.

Climate Change Impacts

Among the effects of climate change are the occurrence of extreme weather events, the elevation of sea levels, and the disturbance of ecosystems. These changes can have cascading effects on agriculture, water resources, and overall environmental stability.

Strategies:

1. **Adaptation Planning:** Develop plans for adapting to changing climate conditions, such as water conservation measures and resilient agricultural practices.

2. **Community Resilience:** Build resilient communities that can collectively address challenges arising from climate change.

3. **Environmental Conservation:** Contribute to environmental conservation efforts to mitigate the long-term impact of climate change.

Nuclear Accidents

Nuclear accidents involve the release of radioactive materials into the environment, often resulting from accidents at nuclear power plants or other facilities.

Strategies:

1. **Emergency Evacuation Plans:** Be aware of evacuation routes and have plans in place in case of a nuclear incident.

2. **Radiation Protection:** Understand and follow guidelines for protection against radiation exposure.

3. **Emergency Shelter:** Identify and prepare emergency shelters that provide protection from radioactive fallout.

Cybersecurity Threats

Cybersecurity threats involve malicious attacks on computer systems, networks, and critical infrastructure, potentially leading to data breaches, financial losses, and disruptions to essential services.

Strategies:

1. **Data Security:** Implement robust cybersecurity measures to protect personal and sensitive information.

2. **Regular Updates:** Keep software and devices updated to patch vulnerabilities.

3. **Education and Awareness:** Stay informed about common cybersecurity threats and educate others to enhance overall awareness.

Geopolitical Tensions

Geopolitical tensions refer to conflicts between nations or regions that can escalate into diplomatic crises, trade disputes, or even military confrontations.

Strategies:

1. **Information Monitoring:** Stay informed about global events and geopolitical developments.

2. **Diplomatic Solutions:** Advocate for diplomatic and peaceful resolutions to conflicts.

3. **Community Cooperation:** Build strong community networks that can provide support during times of geopolitical instability.

LET'S GET YOUR HANDS DIRTY RIGHT AWAY!

Just snap a photo of the QR code—no need to share any personal info—and you'll get direct access to **80 AMAZING projects**! You can print them out on A4 paper, giving you plenty of room to jot down notes, add your own ideas, and really make them your own. It's a simple way to keep your book pristine while customizing these projects to fit your needs perfectly!

ENJOY!

Chapter 5:

Bugging In vs. Bugging Out

Within the domain of emergency readiness and survivalism, two primary strategies frequently take center stage: bugging in and bugging out. These methods embody distinct reactions to diverse crises, and grasping the subtleties of each is essential for adept emergency planning.

Bugging In

Bugging in, also known as sheltering in place, involves staying at home or a designated location during an emergency. This strategy relies on fortifying one's current residence to withstand and endure the challenges posed by a crisis. Here's a detailed breakdown of bugging in:

Advantages of Bugging In:

1. **Familiarity and Comfort:** Bugging in allows individuals to stay in familiar surroundings, offering a sense of comfort and security during challenging times.

2. **Resource Access:** Being at home provides access to pre-established resources, including food stockpiles, water, and emergency supplies.

3. **Community Support:** Neighbors and local community networks can provide mutual support, enhancing overall resilience.

4. **Infrastructure Reliance:** Bugging in relies on existing infrastructure, such as utilities and municipal services, which can be advantageous if these systems remain operational.

5. **Cost-Effectiveness:** Bugging in can be more cost-effective, as it doesn't require immediate relocation or investment in alternative shelters.

Guidelines for Bugging In:

1. **Emergency Supplies:** Ensure you have an ample supply of emergency essentials, including non-perishable food, water, medical supplies, and other necessary items.

2. **Fortification:** Strengthen the security of your home by reinforcing doors and windows, installing security systems, and having a reliable backup power source.

3. **Communication Plans:** Establish clear communication plans with family members and neighbors to share information and coordinate efforts if needed.

4. **Emergency Services Awareness:** Stay informed about local emergency services and community resources, as they can be crucial during crises.

5. **Long-Term Sustainability:** Consider sustainable practices, such as rainwater harvesting and backyard gardening, to enhance long-term self-sufficiency.

Bugging Out

Bugging out involves leaving one's home or current location to seek safety elsewhere, typically in response to a situation where staying is deemed unsafe or untenable. This strategy prioritizes mobility and adaptability. Here's an in-depth look at bugging out:

Advantages of Bugging Out:

1. **Escape from Danger:** Bugging out provides an opportunity to escape from immediate dangers, such as natural disasters, civil unrest, or impending threats.

2. **Versatility:** Bugging out allows for greater adaptability, enabling individuals to respond dynamically to evolving situations.

3. **Strategic Relocation:** A well-planned bug-out strategy involves relocating to a predetermined, safer location, such as a designated bug-out shelter or the homes of friends or family in more secure areas.

4. **Reduced Dependency:** Bugging out reduces reliance on local infrastructure, allowing individuals to access alternative resources and avoid potential shortages.

5. **Survival Skills Utilization:** Bugging out may necessitate the use of survival skills, such as navigation, foraging, and shelter-building, contributing to overall self-sufficiency.

Guidelines for Bugging Out:

1. **Emergency Evacuation Plan:** Develop a clear and well-rehearsed evacuation plan, including multiple routes and alternative destinations based on different scenarios.

2. **Bug-Out Bag:** Prepare a comprehensive bug-out bag containing essential supplies, including clothing, food, water, first aid items, and important documents.

3. **Communication Devices:** Make sure to equip yourself with dependable communication tools like two-way radios or satellite phones to maintain connectivity throughout your bug-out expedition.

4. **Strategic Location Knowledge:** Familiarize yourself with potential bug-out locations and their surroundings, considering factors like accessibility, resources, and security.

5. **Continuous Training:** Regularly practice bug-out scenarios to refine your plan, identify potential challenges, and improve response times.

Considerations for Choosing Between Bugging In and Bugging Out:

1. **Nature of the Threat:** Assess the specific nature of the threat or emergency to determine whether it's safer to stay at home or evacuate.

2. **Available Resources:** Consider the availability of resources at your current location and whether bugging in is sustainable based on your preparedness measures.

3. **Health and Mobility:** Evaluate your health condition and mobility. Bugging out may be challenging for individuals with health issues or limited mobility.

4. **Community Support:** Consider the strength of your community network. Bugging in may be more viable if you have a supportive community, while bugging out may be necessary if community support is limited.

5. **Preparedness Level:** The level of preparedness, including the presence of emergency supplies, a well-thought-out plan, and regular training, will influence the effectiveness of both strategies.

Chapter 6:

Water, Fire, and Food in Survival

Within survival, the essential foundation for enduring and thriving relies on the triumvirate of water, fire, and food. These elements play vital roles in sustaining life, offering comfort, and enhancing resilience during difficult situations.

Water

Water is unequivocally the most crucial element for human survival. The human body is composed of about 60% water, and maintaining proper hydration is paramount for various physiological functions. In survival situations, the availability and access to clean water can be a determining factor for survival. Here's a closer look at the critical role of water and guidelines for managing it effectively:

Role of Water in Survival:

1. **Hydration:** The drinking of a sufficient quantity of water is necessary for the maintenance of biological functions, the regulation of body temp., and the maintenance of overall health.

2. **Nutrient Transport:** Additionally, water acts as a vehicle for the transportation of critical nutrients throughout the body, which in turn makes it easier for the body to absorb minerals and vitamins.

3. **Thermoregulation:** In extreme temperatures, water helps regulate body temperature through processes like sweating and evaporation.

4. **Digestion:** Water is vital for the digestion of food, aiding in the breakdown of nutrients and the absorption of energy.

Guidelines for Water Management in Survival:

1. **Water Sourcing:** Identify and locate nearby water sources, such as rivers, lakes, or streams. Carry maps or GPS devices to navigate to water points.

2. **Purification:** If you want to make sure that the water you drink is safe to consume, you can purify it by boiling it, using water purification pills, or utilizing portable water filters.

3. **Water Conservation:** Practice water conservation by rationing your water supply and avoiding unnecessary water usage.

4. **Storage:** If possible, store extra water in containers or portable water reservoirs for times when access to fresh water may be limited.

5. **Knowledge of Local Flora:** Learn about local plants that may contain water, such as certain types of cacti, to supplement your water sources.

Fire

Fire has been a companion to humanity for millennia, offering warmth, light, and the means to cook food. In survival scenarios, fire serves multiple purposes beyond mere comfort.

It becomes a tool for signaling, protection, and psychological well-being. Understanding the role of fire and mastering fire-making techniques is crucial for survival:

Role of Fire in Survival:

1. **Warmth:** In cold environments, fire provides essential warmth, preventing hypothermia and ensuring the body maintains a suitable temperature.

2. **Cooking:** Fire enables the cooking of food, making it more palatable and aiding in the breakdown of certain toxins or pathogens.

3. **Signal for Rescue:** A well-tended fire can serve as a visible signal for rescue teams, especially in open areas or at night.

4. **Insect Repellent:** Smoke generated by a fire can act as a natural insect repellent, creating a more comfortable living environment.

Guidelines for Fire Management in Survival:

1. **Fire Starting Skills:** Learn various fire-starting techniques, including using matches, lighters, fire starters, and primitive methods like friction fire techniques.

2. **Tinder and Kindling:** Collect tinder, kindling, and fuel wood in advance. Tinder ignites easily, kindling sustains the initial flame, and fuel wood maintains a lasting fire.

3. **Fire Safety:** Follow safety measures to prevent the uncontrolled spread of fire. Clear the area around your fire pit and have a means of extinguishing the fire if needed.

4. **Sustainable Firewood Harvesting:** In a long-term survival situation, practice sustainable harvesting of firewood to ensure the continued availability of this critical resource.

5. **Alternative Fire Sources:** Carry alternative fire-starting tools, such as magnesium fire starters or waterproof matches, to ensure redundancy in your fire-making capabilities.

Food

Food is the fuel that sustains the body's energy levels and supports overall health. In survival scenarios, securing a reliable source of nutrition becomes paramount for long-term well-being and the ability to navigate challenges. Understanding the role of food and implementing effective strategies for acquiring and managing it is vital:

Role of Food in Survival:

1. **Energy:** Food provides the necessary energy to fuel physical activities, maintain body temperature, and support cognitive functions.

2. **Mental Health:** Adequate nutrition is linked to mental well-being. In survival situations, maintaining morale and mental clarity is crucial for decision-making and problem-solving.

3. **Physical Endurance:** A balanced diet contributes to physical strength and endurance, aiding in the completion of tasks necessary for survival.

4. **Immune System Support:** Proper nutrition supports a robust immune system, crucial for resisting diseases and infections in challenging environments.

Guidelines for Food Management in Survival:

1. **Emergency Rations:** Pack emergency food rations, such as energy bars, dried fruits, and nuts, in your survival kit for immediate sustenance.

2. **Foraging:** Learn to identify edible plants, fruits, and berries in your environment. However, exercise caution and only consume plants you can positively identify.

3. **Fishing and Hunting Skills:** If in a location with access to water bodies or wildlife, learn fishing and hunting skills to procure fresh protein sources.

4. **Food Preservation:** For the purpose of extending the shelf life of perishable foods, it is recommended that, in the event of a lengthy survival scenario, food preservation procedures such as drying, smoking, or salting be utilized.

5. **Water-rich Foods:** Consume water-rich foods like fruits and vegetables to supplement your hydration needs.

Integration of Water, Fire, and Food in Survival

While each element—water, fire, and food—plays a distinct role in survival, their integration is key to long-term resilience. Consider the following strategies for effectively managing this trinity in a survival context:

1. **Prioritize Basic Needs:** Address the basic needs of water, fire, and food in that order. Prioritizing water is crucial, as dehydration can quickly lead to a decline in physical and mental functions.

2. **Build Synergies:** Utilize the interplay between these elements. For instance, using fire to purify water or cook food, or strategically placing your campfire to deter insects.

3. **Strategic Campsite Selection:** Choose a campsite that provides access to water and is conducive to fire management. Consider factors like wind direction and the availability of natural resources.

4. **Plan for Redundancy:** Incorporate redundancy into your survival strategy. Have multiple means of water purification, various fire-starting tools, and a diverse range of food sources.

5. **Skills Development:** Continuously enhance your skills in water purification, fire-making, and foraging to ensure adaptability in different survival scenarios.

Chapter 7:

Canning and Food Preservation

anning is a time-honored method of food preservation that allows you to enjoy the bounty of the harvest long after the growing season. Whether you're a seasoned gardener with an abundance of produce or someone looking to reduce food waste, canning is a practical and rewarding skill to acquire. This step-by-step guide will take you through the process of water bath canning, a method suitable for high-acid foods like fruits, pickles, and jams.

Step-by-Step Guide:

Step 1: Gather Your Equipment and Ingredients

Before diving into the canning process, gather all the necessary equipment and ingredients. Here's what you'll need:

Equipment:

1. Water bath canner

2. Canning jars with lids and bands

3. Jar lifter

4. Canning funnel

5. Ladle

6. Bubble remover or non-metallic spatula

7. Clean kitchen towels or paper towels

8. Timer

9. Cutting board and knife

10. Small saucepan for heating jar lids

Ingredients:

1. Fresh fruits or vegetables

2. Acid (lemon juice or vinegar)

3. Sugar (for jams and preserves)

4. Pectin (if making jams or jellies)

5. Pickling salt (for pickles)

Step 2: Sterilize Your Jars

Ensure your jars are thoroughly cleaned before sterilizing. Submerge the jars in simmering water (180°F/82°C) for at least 10 minutes. Keep them in hot water until ready to use.

Step 3: Prepare Your Produce

Wash and prepare your fruits or vegetables according to your chosen recipe. If you're making pickles, blanch vegetables like cucumbers briefly in boiling water to enhance crispness.

Step 4: Fill the Water Bath Canner

The water bath canner should be filled with sufficient water to cover the jars by a degree that is at least one to two inches. The water should be brought to a simmer. It's important to start with hot water to minimize temperature shock when the jars are added.

Step 5: Prepare the Canning Syrup or Liquid

If your recipe requires a syrup or liquid, prepare it according to the guidelines. This may involve dissolving sugar in water or creating a pickling solution with vinegar, salt, and spices.

Step 6: Fill the Jars

Place the canning funnel on top of a sterilized jar. Carefully pack the prepared produce into the jars, leaving the recommended headspace (the space between the top of the food and the rim of the jar) specified in your recipe.

Step 7: Add Acid

To ensure safe acidity levels, add lemon juice or vinegar to each jar. This is particularly important for fruits and tomatoes. Consult your recipe for the correct amount.

Step 8: Pour in the Syrup or Liquid

If your recipe calls for a syrup or liquid, carefully pour it over the produce in the jar. Leave the specified headspace.

Step 9: Remove Air Bubbles

Insert the bubble remover or non-metallic spatula into the jar to release any trapped air bubbles. This step helps prevent spoilage and ensures proper sealing.

Step 10: Wipe Jar Rims and Attach Lids

Take a fresh, moist cloth or paper towel to wipe the edges of the jars, eliminating any leftover material that might hinder proper sealing. Position the sanitized lids onto the jars and fasten them using the bands, tightening to fingertip-tightness.

Step 11: Process in the Water Bath Canner

Using the jar lifter, carefully lower the filled and lidded jars into the simmering water in the canner. Ensure the jars are fully submerged, with water covering them by at least 1 to 2 inches. Bring the water to a rolling boil.

Step 12: Process for the Recommended Time

Start the timer once the water reaches a rolling boil. Process the jars for the recommended time specified in your recipe. This processing time ensures that harmful microorganisms are destroyed, and a vacuum seal is created.

Step 13: Remove Jars and Cool

Once the processing is complete, deactivate the heat source and employ the jar lifter to take out the jars from the canner. Put them on a clean kitchen towel or cooling rack, leaving space between each jar for air circulation.

Step 14: Listen for Sealing Pops

As the jars cool, you may hear the distinctive "pop" sound of the lids sealing. This indicates a successful vacuum seal. Allow the jars to cool completely, typically for 12 to 24 hours.

Step 15: Check Seals and Store

Once the jars are fully cooled, press down on the center of each lid. If it doesn't pop back, the jar is sealed. If any jars haven't sealed, refrigerate or reprocess the contents promptly. Keep sealed jars in a cool, dim, and dry environment.

Tips for Successful Canning:

1. **Follow Tested Recipes:** Use recipes from reputable sources, such as the USDA or Ball Canning, to ensure safe processing times and ingredient proportions.

2. **Use High-Quality Produce:** Choose fresh, high-quality produce for the best results. Avoid overripe or underripe fruits and vegetables.

3. **Be Mindful of Altitude:** Adjust processing times for your altitude. Higher altitudes may require longer processing times.

4. **Label and Date Jars:** Ensure that you clearly mark each jar with its contents then the date it was canned. This helps you keep track of freshness and enables you to rotate your preserved goods effectively.

5. **Inspect for Spoilage:** Before consuming canned items, examine the jars for indications of spoilage, such as lids that are swollen, unpleasant odors, or unusual discoloration.

By following these step-by-step instructions and adhering to safe canning practices, you can confidently preserve the flavors of the season and build a pantry stocked with your own homemade jams, pickles, and other delicious treats. Canning not only offers a means of reducing food waste but also provides a satisfying connection to traditional food preservation methods that have stood the test of time.

Chapter 8:

Mental Health in Crises

During challenging periods like natural disasters, pandemics, or personal hardships, prioritizing mental health becomes essential. Managing stress, uncertainty, and adversity significantly contributes to overall well-being.

Understanding the Impact of Crises on Mental Health

Crises can exert a profound impact on mental health, triggering a range of emotional responses such as fear, anxiety, sadness, and even trauma. The uncertainty and disruption that crises bring can overwhelm individuals, leading to heightened stress levels. Acknowledging the potential toll on mental health is the first step in developing effective strategies for coping.

Strategies for Stress Management:

1. **Mindfulness and Relaxation Techniques**

Deep Breathing: Employ deep breathing techniques to soothe the nervous system. Inhale profoundly, retain your breath for a few moments, and exhale gradually.

Mindfulness Meditation: Participate in mindfulness meditation to center your attention on the current moment. Utilize apps and online tools to assist you in navigating through meditation sessions.

2. Physical Activity

Exercise Regularly: Make physical activity a consistent part of your schedule to effectively alleviate stress. Whether it's walking, jogging, practicing yoga, or doing home workouts, regular exercise can be a powerful stress reliever.

Outdoor Activities: Allocate time to connect with nature, as being outdoors can significantly enhance your mood and diminish stress levels.

3. Maintaining a Routine

Create Structure: Establish a daily routine to provide a sense of normalcy and control. Structure can contribute to stability during uncertain times.

Prioritize Self-Care: Include self-care activities in your routine, such as adequate sleep, healthy meals, and relaxation breaks.

4. Social Connection

Stay Connected: Maintain social connections, even if virtually. Consistent interaction with friends and family can offer emotional assistance and alleviate sentiments of loneliness.

Join Support Groups: Participate in online or local support groups where individuals share similar experiences. Finding solace in the company of those who comprehend your struggles can be reassuring.

5. Limit Information Consumption

Set Boundaries: While staying informed is important, set boundaries on media consumption. Overconsumption of news and social media can lead to heightened stress and anxiety levels.

Fact-Check Information: Verify information from reliable sources to avoid misinformation that may contribute to heightened stress levels.

6. **Expressing Emotions**

Journaling: Keep a journal to document your thoughts and emotions, offering a constructive way to express feelings and gain valuable perspective..

Talk to a Professional: Consider consulting a mental health professional, like a counselor or therapist, for support in discussing and navigating your emotions.

Building Resilience

1. **Cultivate a Positive Mindset**

Focus on Strengths: Identify your strengths and past experiences of overcoming challenges. Cultivate a positive mindset by acknowledging your resilience.

Practice Gratitude: Consistently show appreciation for the positive elements in your life. Doing so can redirect your attention to the optimistic aspects, even in the face of difficulties.

2. **Adaptability**

Embrace Change: Develop a mindset that embraces change as a natural part of life. Being adaptable allows for a smoother adjustment to unexpected circumstances.

Problem-Solving Skills: Improve your ability to solve problems so that you can tackle obstacles with a mindset that is focused on finding solutions.

3. **Self-Reflection**

Learn from Experiences: Reflect on past challenges and crises to identify lessons learned. Understanding how you've coped before can inform your approach to current situations.

Personal Growth: View challenges as opportunities for personal growth. Each crisis can provide a chance to develop resilience and new coping strategies.

4. **Community Support**

Build a Supportive Network: Strengthen connections with friends, family, and community members. A supportive network can provide both practical assistance and emotional support.

Offer Support: Offering assistance to fellow community members nurtures a feeling of collective responsibility and interconnectedness.

5. **Set Realistic Goals**

Break Tasks into Smaller Steps: When facing overwhelming challenges, break tasks into smaller, more manageable steps. This approach can make the path forward clearer and less daunting.

Celebrate Achievements: Acknowledge and celebrate small victories along the way. Recognizing achievements contributes to a sense of accomplishment.

6. **Coping Strategies**

Healthy Coping Mechanisms: Identify and cultivate healthy coping mechanisms. This might involve pursuing hobbies, dedicating time to cherished ones, or participating in activities that evoke happiness.

Avoid Negative Coping: Be mindful of negative coping mechanisms such as excessive alcohol consumption or unhealthy behaviors. Seek alternatives that contribute to well-being.

LET'S MOVE FROM WORDS TO ACTION!

SCAN ME / SCAN ME / SCAN ME / SCAN ME / SCAN ME / SCAN ME / SCAN ME

DOWNLOAD
THE RESILIENCE DIARY

Chapter 9:

Shelter Essentials

Survival and outdoor activities demand a fundamental need for shelter. Whether navigating the wilderness, encountering emergencies, or preparing for a camping excursion, a deep grasp of shelter essentials proves indispensable.

Types of Shelters

Tents
- **Dome Tents:** Easy to set up and relatively stable, dome tents are popular for camping and backpacking. They come in various sizes, accommodating different numbers of occupants.
- **A-frame Tents:** Simple and traditional, A-frame tents are known for their stability. They are suitable for various environments and are often used in camping scenarios.
- **Tunnel Tents:** Long and tunnel-shaped, these tents are lightweight and offer ample interior space. They are suitable for backpacking and camping trips.

Tarps and Hammocks
- **Tarp Shelters:** Lightweight and versatile, tarps can be configured into various shelter styles, such as A-frames, lean-tos, or even improvised tents. They are popular among minimalist campers and survival enthusiasts.
- **Hammock Shelters:** Combining a hammock with a rainfly or tarp provides an elevated and comfortable shelter option, particularly in areas with challenging terrain.

Emergency Shelters
- **Space Blankets:** Compact and reflective, space blankets are designed for emergency situations. They reflect body heat and can be used as improvised shelters or added insulation to other shelters.
- **Emergency Bivvy Bags:** These compact, lightweight bags provide a quick and effective way to create an emergency shelter. They are often made from reflective materials for added insulation.

Natural Shelters
- **Caves and Rock Overhangs:** Natural features such as caves or rock overhangs can provide ready-made shelters. However, ensure safety by checking for stability and potential hazards.
- **Debris Huts:** Constructed using natural materials like branches, leaves, and debris, these huts provide insulation and protection. They are suitable for survival situations.

Snow Shelters
- **Snow Caves:** Dug into a snowbank or drift, snow caves provide insulation and protection from the elements. They are commonly used in snowy environments for winter survival.
- **Quinzee Shelters:** Constructed by piling and compacting snow, quinzees offer a viable option for shelter in snowy conditions. They provide insulation against the cold.

Shelter Construction Basics

Site Selection
- **Level Ground:** Choose a level ground to avoid discomfort during sleep and ensure proper stability for your shelter.
- **Avoid Hazards:** Check for potential hazards such as falling branches, uneven terrain, or areas prone to flooding.

Materials
- **Natural Materials:** If constructing a shelter using natural materials, gather sturdy branches, leaves, and debris. Ensure they are free from insects and other potential hazards.
- **Tarp and Cordage:** For tarp or hammock shelters, have a reliable tarp or rainfly and suitable cordage for securing the shelter.

Anchoring and Stability
- **Sturdy Anchors:** Ensure that your shelter is securely anchored. For tents, use stakes or anchors provided with the tent. For improvised shelters, use heavy rocks or logs.
- **Wind Direction:** Consider the direction of prevailing winds when setting up your shelter to minimize exposure.

Insulation
- **Ground Insulation:** Lay a groundsheet or insulating material beneath your shelter to prevent heat loss to the ground.
- **Layering:** In cold conditions, use layers of natural materials or additional tarps to provide insulation and wind protection.

Ventilation
- **Avoid Condensation:** Ensure proper ventilation to prevent condensation inside your shelter, especially in colder weather. Leave openings or vents where necessary.

Situational Appropriateness

Camping Trips
- **Tents:** For camping trips, traditional tents, dome tents, or tunnel tents are excellent choices. They offer convenience, comfort, and protection from the elements.

Backpacking
- **Lightweight Options:** Consider lightweight options such as backpacking tents, tarps, or hammock shelters. These provide portability without compromising protection.

Survival Scenarios
- **Emergency Shelters:** In survival situations, prioritize compact and easily deployable options such as space blankets, bivvy bags, or improvised debris huts.

Winter Survival
- **Snow Shelters:** In snowy environments, snow caves, quinzees, or properly insulated tents with cold-weather ratings are essential for winter survival.

Emergency Situations
- **Quick Deployment:** In emergencies, prioritize shelters that can be set up quickly and easily, such as space blankets, emergency bivvy bags, or improvised natural shelters.

Outdoor Exploration
- **Versatility:** For versatile outdoor exploration, choose shelters like tarps or hammocks that allow adaptation to various environments and conditions.

BOOK 2:
WORST CASE SCENARIO
PRACTICAL GUIDE

Chapter 1:

Surviving a Nuclear Disaster

When the world faces the ominous specter of a nuclear disaster, envisioning the worst-case scenario is crucial. Picture a scenario where the air is thick with radioactive fallout, and the once bustling landscape turns into a desolate, hazardous zone. In this grim setting, survival becomes a primal instinct, requiring quick thinking and practical strategies.

Survival Strategies:

1. **Shelter Construction**: In the aftermath of a nuclear disaster, shelter is your first line of defense. With basic materials at your disposal, learn how to construct makeshift shelters that can shield you from the harmful effects of radioactive fallout. Whether it's finding refuge in buildings or crafting simple structures, the key is to create a protective barrier against radiation.

2. **Radiation Monitoring**: Understanding radiation is vital for survival. Arm yourself with knowledge on basic radiation principles, and then learn how to monitor radiation levels using improvised tools. Being aware of your surroundings and being able to gauge radiation intensity is crucial for making informed decisions about your safety.

3. **Essential Supplies**: Surviving in a post-nuclear world requires a well-equipped survival kit. Compile a detailed list of essential supplies, including protective clothing to shield yourself from radiation exposure, methods for purifying water to ensure a clean water supply, and options for detoxifying your body from radiation. These supplies form the backbone of your survival toolkit.

4. **Navigating Contaminated Areas**: In a world tainted by radioactive fallout, knowing how to navigate contaminated areas is a skill that can save your life. Gain insights into safe navigation techniques that minimize your exposure to radiation. Whether it's identifying safer routes or utilizing protective gear, understanding how to move through contaminated areas is essential for survival.

Chapter 2:

Navigating a Pandemic Crisis

In the face of a global viral outbreak, the worst-case scenario unfolds as communities grapple with the relentless spread of illness, overwhelming healthcare systems, and a pervasive sense of uncertainty. Families find themselves isolated, economies falter, and daily life comes to a standstill as the pandemic wreaks havoc.

Survival becomes paramount in such dire circumstances, and adopting effective strategies is key to weathering the storm. Here are practical survival tips:

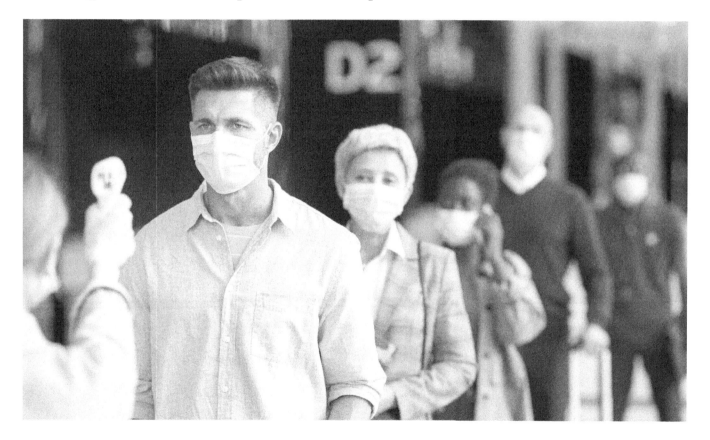

1. **Quarantine Protocols:** Learn how to establish effective quarantine measures to protect yourself and your community.

2. **Medical Readiness**: Arm yourself with essential medical supplies and knowledge. Compile a list of basic medical necessities such as bandages, pain relievers, and fever reducers. Know basic first aid techniques to handle emergencies at home.

3. **Building Immunity:** Understand the role of nutrition, exercise, and stress management in boosting your immune system.

4. **Communication Connectivity**: In times of crisis, communication is a lifeline. Establish reliable communication channels with neighbors, friends, and family. Stay informed about the latest developments, share valuable resources, and offer support to those in need. Social connectivity can be a powerful tool in navigating the challenges of a pandemic.

Chapter 3:

Surviving Natural Disasters

Nature, with all its grandeur, can sometimes unleash its fury in the form of natural disasters. These events, ranging from earthquakes to hurricanes, pose significant threats to human safety and well-being. In the worst-case scenario, individuals may find themselves grappling with the aftermath of a devastating event, facing challenges such as damaged homes, disrupted infrastructure, and scarcity of essential resources.

Surviving such natural disasters requires strategic planning and resilience. Here are practical survival strategies to navigate through the chaos and emerge stronger:

1. **Earthquake Preparedness:** The ground shaking beneath your feet can be terrifying, but being prepared can make a difference. Secure furniture and heavy objects in your living space to minimize hazards. Putting together emergency kits that include things like water, food that does not go bad, materials for first aid, and critical documents is a wonderful idea. Additionally, educate yourself on the actions to take during and after an earthquake to ensure your safety.

2. **Hurricane Survival:** Hurricanes unleash formidable winds and heavy rainfall, posing a threat to homes and communities. Preplan evacuation routes and identify a secure shelter. Strengthen your residence by securing doors and windows, and ensure you have essential provisions like canned food, batteries, and medications. Exercise caution during the aftermath of the storm, as potential hazards may persist.

3. **Wildfire Safety:** In the face of a raging wildfire, understanding its behavior is crucial. Create defensible zones around your property by clearing away flammable vegetation. Have an evacuation plan in place, including a designated meeting point for family members. Keep emergency supplies ready to go, including clothing, important documents, and a communication plan to stay connected with loved ones.

4. **Flood Resilience:** Flooding can be both sudden and destructive. Implement flood-resistant measures for your home, such as elevating electrical systems and installing barriers to

prevent water entry. Know the evacuation routes in your area and have a plan in place to reach higher ground. Ensure access to emergency supplies, including a sufficient quantity of clean water, non-perishable food, and necessary medications.

Chapter 4:

Surviving Power Grid Failures

Picture a scenario where the power grid suddenly collapses, plunging entire communities into darkness. No lights, no appliances, and no gadgets. The worst-case scenario unfolds as chaos ensues, with essential services disrupted, leaving people vulnerable and uncertain about the future.

Survival Strategies:

1. **Off-Grid Living:** When the lights go out, knowing how to live off the grid becomes crucial. Understand alternative energy sources like solar panels and wind turbines. Learn to manage water resources efficiently, from harvesting rainwater to purifying it for consumption. Master the art of food storage, preserving essentials without relying on refrigeration.

2. **Communication in Isolation:** With traditional communication channels down, staying connected becomes a lifeline. Establish reliable methods like two-way radios or ham radios. Create a communication plan with family and neighbors, ensuring everyone knows how to reach each other when smartphones and the internet are no longer viable.

3. **Security Measures:** In the absence of conventional law enforcement, securing your home and community becomes paramount. Explore strategies such as neighborhood watch programs, setting up community patrols, and fortifying your living space. Basic security measures can deter potential threats and create a sense of safety within your community.

4. **Essential Skills:** In a world without power, practical skills become invaluable. Learn the basics of gardening to grow your own food, reducing dependence on external sources. Develop basic carpentry skills for repairs and construction. Acquire first aid knowledge to handle medical emergencies, as hospitals may be inaccessible. These essential skills enhance self-sufficiency, allowing you to navigate daily life without the usual conveniences.

Chapter 5:

Coping with Urban Disruptions

L iving in crowded cities can become a challenge when faced with unexpected disruptions. Picture a scenario where the city is engulfed in chaos, from civil unrest to breakdowns in the supply chain. The worst-case situation involves being caught in the middle of a gridlock nightmare, with resources dwindling and the usual order of urban life collapsing.

Surviving in such an environment requires strategic thinking and preparedness. Here's a practical guide to dealing with the chaos and coming out unscathed:

Survival Strategies:

1. **Urban Escape Routes:** Imagine the roads clogged with traffic, making movement nearly impossible. Plan ahead for safe evacuation routes, considering alternate paths that may be less congested. Familiarize yourself with the city's layout, ensuring you have multiple options to reach a safer location.

2. **Community Bonds:** In times of crisis, the strength of community support is invaluable. Build connections with your neighbors and fellow residents, creating a network that can share resources and provide emotional support. A united community is more resilient and better equipped to face the challenges of urban disruptions.

3. **Resource Mastery:** As the chaos unfolds, resources like water, food, and medical supplies become scarce. Learn essential skills for managing these limited resources efficiently. Know where to find alternative water sources, stockpile non-perishable food items, and create a basic medical kit to handle emergencies.

4. **Urban Survival Skills:** Urban survival requires specific skills tailored to the cityscape. Familiarize yourself with navigating public transportation disruptions, understanding the subway or bus systems, and identifying safe zones. Secure your residence by fortifying entry points and having a plan for unexpected power outages.

BOOK 3:
HOME DEFENSE GUIDE

Chapter 1:

Securing the Home

Ensuring the safety and protection of your home is a crucial element in creating a secure environment for you and your loved ones. By applying successful methods to strengthen your home, you improve its ability to withstand potential risks. Here, we delve into practical approaches that center on protecting your living space.

1. Reinforcing Entry Points

Start by reinforcing the entry points of your home, such as doors and windows. Install sturdy locks and deadbolts, ensuring they meet industry standards for security. Consider upgrading to smart locks for added convenience and control. Reinforce door frames with metal plates to make them more resistant to forced entry.

For windows, use laminated glass or install security film to make them shatter-resistant. Window bars or grilles can act as a physical barrier, especially for ground-level windows. Assess the vulnerability of sliding doors and install secondary locks or bars to prevent them from being easily forced open.

2. Fortifying Perimeters

Create a strong perimeter around your home to act as an initial deterrent. Install fences or walls with secure gates to control access points. Ensure that the fence is sturdy and not easily scalable. Plant thorny bushes or shrubs near windows to discourage unauthorized access.

Boost your security effortlessly with outdoor lighting. Place motion-activated lights along the outer edges of your residence. Adequately illuminated surroundings discourage potential intruders and facilitate the detection of any unusual activities.

3. Security Systems and Surveillance

Invest in modern security systems to add an extra layer of protection. Install a comprehensive security alarm system that covers all entry points. In this category are included sensors for doors and windows, motion detectors, and sensors for glass breakage. Ensure that the alarm is monitored, either professionally or by a reliable DIY system.

Surveillance cameras provide valuable visual information and act as a deterrent. Install cameras at strategic locations, covering entry points and vulnerable areas. Choose high-quality cameras equipped with night vision features for continuous 24/7 surveillance.

4. Securing Doors

Doors are primary entry points, and securing them is crucial. Install solid-core doors that are more resistant to forced entry. Reinforce the door frame with metal plates and longer screws to enhance stability. Consider installing a peephole or a smart doorbell camera for visual verification of visitors.

For sliding glass doors, place a sturdy rod or bar in the track to prevent forced opening. Consider adding security film to the glass for additional protection. Patio doors can be vulnerable, so reinforce them with secondary locks or bars.

5. Safe Room Preparation

Create a designated safe room within your home for emergencies. Choose a room with minimal windows, preferably with a solid-core door. Reinforce the door and frame, and equip the room with essential supplies, including a first aid kit, emergency communication devices, and any necessary tools for self-defense.

Consider installing a safe or lockbox within the safe room to secure important documents, valuables, and additional supplies. Ensure that all family members are aware of the safe room's location and its purpose in case of an emergency.

6. Exterior Maintenance

Ensuring the regular maintenance of your home's exterior is essential for maintaining security. Prune bushes and trees close to windows to eliminate potential hiding spots for intruders. Keep the landscaping well-maintained to ensure clear visibility around your property.

Maintain the exterior of your home to prevent it from appearing vacant. Repair any broken windows promptly, and address any structural issues that may compromise the security of your home.

7. Neighbor Collaboration

Build strong relationships with your neighbors and foster a sense of community. Collaborate with neighbors on neighborhood watch initiatives to enhance overall security. Share information about any suspicious activity, and encourage open communication.

Consider establishing a reciprocal agreement with neighbors to keep an eye on each other's properties when one is away. This collaborative approach creates a supportive network that adds an extra layer of vigilance to the security of your home.

8. Home Automation for Security

Explore home automation options to increase the efficiency of your security measures. Smart home devices enable you to conveniently oversee and regulate different facets of your home security. This encompasses intelligent doorbell cameras, smart locks, and home security systems that can be administered via mobile applications.

Automation can also simulate occupancy when you're away. Smart lighting and programmable thermostats can be set to create the appearance that someone is home, deterring potential burglars.

9. Emergency Preparedness

Incorporate emergency preparedness into your home security plan. Have a well-thought-out evacuation plan and ensure that all family members are familiar with it. Keep emergency supplies, including food, water, and essential medications, readily accessible.

Regularly update and rehearse your emergency plan to ensure everyone knows what to do in different scenarios. Consider conducting periodic drills to practice evacuation and communication procedures.

10. Awareness and Mindfulness

Stay vigilant and be aware of your surroundings. Mindful living involves being observant of any changes in your neighborhood and staying informed about local security concerns. Stay updated on local crime trends and adjust your security measures accordingly.

Educate family members about security awareness, emphasizing the importance of locking doors and windows, even when at home. Encourage a culture of security consciousness within your household.

Chapter 2:

Perimeter Defense

Securing the perimeter of your property is a foundational step in creating a safe and protected environment. By implementing effective strategies for perimeter defense, you not only establish a deterrent against potential threats but also fortify the first line of defense for your home. Let's explore practical and actionable techniques to safeguard your property boundaries.

1. **Fencing Solutions**

A sturdy and well-designed fence is a fundamental aspect of perimeter defense. Choose a fence that suits both your security needs and complements the aesthetics of your property. Consider options such as:

- **Metal Fencing:** Steel or wrought iron fences are durable and provide a formidable barrier. They can be designed with pointed or spiked tops for added security.

- **Wooden Fencing:** While offering a more traditional look, wooden fences can be robust and provide privacy. Opt for solid panels and ensure they are well-maintained to prevent vulnerabilities.
- **Chain-Link Fencing:** While not as visually imposing, chain-link fences are cost-effective and can be enhanced with privacy slats. They offer visibility, allowing you to see potential threats.

2. **Access Control**

Controlling access points is crucial for perimeter defense. Limit the number of entry points to your property and focus on securing them effectively. Implement access control measures such as:

- **Gates and Barriers:** Install secure gates at driveways and entry points. Consider automated gates with access control systems, allowing you to regulate entry through codes, cards, or remote controls.
- **Security Checkpoints:** Implement security checkpoints, where necessary, to verify individuals before granting access. This could include intercom systems or video surveillance to identify visitors.

3. **Natural Barriers**

Incorporate natural elements into your perimeter defense strategy to act as additional barriers. Nature can be a powerful ally in creating a deterrent. Consider:

- **Landscaping:** Plant dense shrubs, thorny bushes, or hedges along the perimeter. These not only create a physical barrier but also limit visibility for potential intruders.
- **Rock Features:** Strategically place large rocks or boulders to impede vehicular access and create a natural obstacle. This can enhance the overall security of your property.

4. **Lighting**

Proper lighting is a cost-effective and impactful strategy for perimeter defense. Well-lit areas deter potential intruders and provide visibility for surveillance. Consider:

- **Motion-Activated Lights:** Install motion-activated lights along the perimeter to illuminate specific areas when motion is detected. This surprises potential intruders and draws attention to their presence.
- **Perimeter Lighting:** Ensure consistent lighting along the entire perimeter, creating a well-lit barrier. This not only enhances security but also contributes to the overall aesthetics of your property.

5. **Surveillance Systems**

Modern surveillance technology plays a crucial role in perimeter defense. Implementing surveillance systems enhances your ability to monitor and respond to potential threats. Key elements include:

- **Cameras:** Install high-quality cameras at strategic locations along the perimeter. Ensure coverage of entry points, fence lines, and vulnerable areas. Choose cameras with night vision capabilities for 24/7 surveillance.
- **Monitoring Stations:** Establish monitoring stations within your home, equipped with screens displaying real-time footage from surveillance cameras. This allows you to stay informed about activities along the perimeter.

6. **Security Signage**

Visible security signage serves both as a deterrent and a warning. Clearly display signs indicating the presence of security measures. This includes signs for:

- **Surveillance Cameras:** Inform potential intruders that they are being monitored. Visible cameras act as a strong deterrent.
- **Private Property:** Clearly mark your property boundaries with signs indicating that the area is private property. This serves as a warning against unauthorized entry.

7. **Patrols and Vigilance**

Active vigilance is a crucial component of perimeter defense. Regular patrols and vigilant observation contribute to a proactive security approach. Consider:

- **Security Personnel:** If feasible, hire security personnel for regular patrols along the perimeter. Their presence can act as a strong deterrent.
- **Community Watch:** Establish or participate in a community watch program. Work together with your neighbors to monitor one another's properties and notify authorities of any unusual or suspicious activities.

8. **Reinforced Entry Points**

Focus on reinforcing specific entry points that may be vulnerable to breaches. This includes:

- **Doors and Windows:** Ensure that doors and windows facing the perimeter are reinforced with security features. Install sturdy locks, reinforce frames, and consider impact-resistant glass.

- **Garage Security:** If applicable, pay special attention to garage doors. These are often overlooked but can serve as potential entry points.

9. **Response Planning**

Prepare for potential security breaches with a well-thought-out response plan. This involves:

- **Emergency Contacts:** Keep a record of essential contacts for emergencies, including local law enforcement and neighbors. Make sure everyone in your household is familiar with these contacts.
- **Communication Protocols:** Establish clear communication protocols within your household. This includes a system for alerting others in case of a security breach and communicating with law enforcement.

10. **Strategic Signaling**

Consider strategic signaling methods to alert you to potential threats. This can include:

- **Sensor Alarms:** Install perimeter sensors that trigger alarms when breached. These can be connected to your overall security system.
- **Audible Deterrents:** Use audible deterrents, such as loud alarms or sirens, to alert you and deter intruders in case of a breach.

Chapter 3:

Home Hardening

Ensuring the safety and security of your home involves a proactive approach known as "home hardening." This process entails reinforcing your home against potential threats, making it a more resilient and fortified environment. Here, we explore practical and straightforward methods for home hardening to enhance the safety of your living space.

1. **Strengthening Entry Points**

Start by focusing on the primary entry points of your home, such as doors and windows. Reinforce exterior doors with sturdy materials and install deadbolt locks for added security. Consider upgrading to smart locks for convenient access control.

Windows are vulnerable points, so reinforce them with laminated glass or install security film. Window bars or grilles act as a physical barrier, deterring break-ins. Use sliding window locks to prevent them from being easily forced open.

2. **Window Coverings for Privacy**

Maintain privacy by using window coverings, such as blinds or curtains. This not only prevents outsiders from peering into your home but also conceals valuables from potential intruders. Choose coverings that can be easily adjusted to allow natural light during the day while providing privacy at night.

3. **Landscaping Strategies**

Landscaping can contribute to both aesthetics and security. Prune shrubs and trees around windows to remove potential hiding places for intruders. Consider planting thorny bushes or shrubs along the perimeter, acting as a natural deterrent.

Ensure clear visibility around your property by keeping the landscaping well-maintained. An unkempt exterior can create hiding spots and obscure potential threats.

4. **Exterior Lighting**

Proper exterior lighting is a simple yet effective method for home hardening. Well-lit surroundings deter potential intruders and improve overall visibility. Implement the following lighting strategies:

- Install motion-activated lights at entry points and along the perimeter.
- Use consistent lighting along pathways and around the entire exterior.
- Consider solar-powered lights for cost-effective and environmentally friendly solutions.

5. **Reinforce Garage Security**

Garages often serve as entry points for intruders. Reinforce garage security by following these steps:

- Install a sturdy garage door with proper locks.
- Use frosted or tinted windows on garage doors to obscure the view inside.
- Secure garage doors with additional locks or deadbolts.

6. **Secure Outdoor Valuables**

Outdoor valuables, such as bicycles, tools, and equipment, are attractive targets for thieves. Secure these items by:

- Using strong locks and chains to tether valuable items.
- Storing tools and equipment in locked sheds or storage units.
- Keeping outdoor furniture and decorations secure to prevent theft or vandalism.

7. **Home Security Systems**

Modern technology offers advanced home security systems to enhance protection. Install a comprehensive security system that includes:

- Surveillance cameras strategically placed around the exterior.
- Motion detectors that trigger alarms in case of unusual activity.
- Window and door sensors connected to the security system.

8. **Reinforce Walls and Windows**

Consider reinforcing the structural elements of your home to withstand potential threats. This includes:

- Impact-Resistant Windows: Upgrade to windows designed to withstand impacts, providing an additional layer of protection.

- Reinforced Exterior Walls: Explore options for reinforcing exterior walls to resist forced entry attempts.

9. **Fire Safety Measures**

Home hardening also involves preparing for potential emergencies, such as fires. Implement fire safety measures:

- Place smoke detectors strategically in important areas throughout your residence.
- Keep fire extinguishers accessible and ensure everyone in your household knows how to use them.
- Have an emergency evacuation plan in place.

10. **Reinforced Safe Room**

Consider designating a safe room within your home for extreme situations. Reinforce this room with:

- Solid-core doors and reinforced door frames.
- Additional locks or deadbolts for added security.
- Essential emergency provisions such as a first aid kit, water, and non-perishable food items.

11. **Security Signage**

Visible security signage acts as a deterrent and sends a clear message to potential intruders. Present signs that denote the existence of security measures, including:

- Surveillance cameras.
- Alarm systems.
- Warning signs for private property.

12. **Communication Preparedness**

Ensure that communication is maintained in case of emergencies or power outages. Establish communication protocols, including:

- Emergency contact lists for family members and neighbors.
- Other communication options, like two-way radios or a specified meeting location.

13. **Regular Maintenance**

Regular maintenance is a key aspect of home hardening. Conduct routine checks to identify and address potential vulnerabilities:

- Examine doors and windows for indications of wear or damage.
- Check the functionality of security systems, alarms, and cameras.
- Ensure exterior lighting is operational and replace any burnt-out bulbs promptly.

14. **Neighborhood Collaboration**

Collaborating with neighbors enhances overall home security. Foster a sense of community by:

- Participating in neighborhood watch programs.
- Sharing information about recent security incidents or suspicious activity.
- Collaborating on security initiatives, such as shared surveillance or lighting solutions.

15. **Home Hardening Education**

Educate your household members about the importance of home hardening. Ensure everyone is aware of security protocols, emergency evacuation plans, and how to use security features. This shared knowledge contributes to a collective effort in maintaining home safety.

Chapter 4:

Secret Storage Solutions

Creating hidden compartments for secret storage is a practical and ingenious way to enhance the security and organization of your living space. Whether you want to safeguard valuables, maintain privacy, or simply declutter, hidden storage solutions offer a discreet and efficient answer to various needs. Let's explore the methods and motivations behind incorporating secret storage into your home.

1. **Concealed Furniture**

One of the most seamless ways to integrate hidden storage is through furniture with concealed compartments. Consider:

- **Secret Drawers:** Furniture pieces like nightstands, coffee tables, or dressers can have hidden drawers that blend seamlessly with the overall design.
- **Lift-Top Tables:** Coffee tables or end tables with lift-top mechanisms reveal hidden storage space underneath, providing a discreet spot for items like remotes or magazines.

2. **False Bottoms and Compartments**

Install false bottoms or hidden compartments within existing furniture or fixtures:

- **False Bottom Drawers:** Create a false bottom within a drawer, leaving a concealed space beneath the visible items.
- **Hidden Shelves:** Install shelves with secret compartments behind them, camouflaged by decorative elements.

3. **Wall-Mounted Concealment**

Utilize wall space for secret storage solutions:

- **Floating Shelves with Hidden Compartments:** Install floating shelves with hidden compartments, accessible by removing a discreet cover.
- **Wall Safes:** Consider wall safes that can be concealed behind artwork or a hinged panel for secure storage of important documents or valuables.

4. **Concealed Doors and Panels**

Integrate secret storage into the very structure of your home:

- **Hidden Door Storage:** Incorporate shelves or compartments into the design of doors. This could include bookshelves that double as doors to hidden rooms or storage spaces.
- **Recessed Wall Panels:** Create recessed wall panels that slide or pivot to reveal hidden storage behind them.

5. **Furniture with Dual Functions**

Opt for furniture pieces that serve dual purposes:

- **Storage Ottomans and Benches:** Furniture that doubles as storage is ideal for concealing items while providing additional seating or functionality.
- **Beds with Storage:** Beds with built-in drawers or storage compartments underneath maximize space while keeping belongings out of sight.

6. **Underfloor Storage**

Utilize underfloor space for hidden storage:

- **Floor Safes:** Install floor safes that are discreetly covered with removable panels or floor coverings.
- **Trapdoors:** Create trapdoors in the floor that lead to concealed storage areas.

7. **Built-In Cabinetry**

Custom-built cabinetry can be designed to include secret compartments:

- **Cabinet False Backs:** Install false backs in cabinets that lead to concealed storage spaces.
- **Pull-Out Pantries:** Cabinets with pull-out pantry shelves can hide items behind the facade of regular cabinet doors.

8. **Secret Storage in Plain Sight**

Hide items in plain sight with clever concealment methods:

- **Book Safes:** Use hollowed-out books to create secret compartments on bookshelves.
- **Clever Containers:** Disguise storage as everyday items, such as tissue boxes, household products, or even fake electrical outlets.

9. **Staircase Storage**

If your home has stairs, consider incorporating hidden storage within the staircase:

- **Under-Stair Storage:** Create drawers or cabinets beneath the stairs to utilize often-underutilized space.
- **Staircase Drawers:** Design staircase steps with built-in drawers for discreet storage.

10. **DIY Concealment Projects**

Explore various do-it-yourself (DIY) projects for hidden storage:

- **Floating Picture Frames:** Turn a picture frame into a hidden compartment by attaching it to the wall with hinges.
- **Hidden Key Storage:** Use everyday objects like fake rocks, outdoor fixtures, or birdhouses to conceal spare keys.

Chapter 5:

Protecting Vulnerable Family Members

Safeguarding the safety and well-being of vulnerable family members, including children and the elderly, necessitates a personalized and considerate approach. Adapting strategies to tackle the distinct needs and challenges encountered by these individuals is crucial for establishing a secure home environment. Let's delve into pragmatic and individualized methods for safeguarding children and the elderly in your household.

Child Safety

Children, by nature, are curious and explorative. Crafting a safe environment for them involves a combination of physical safeguards, education, and constant supervision.

- **Childproofing:** Introduce child safety precautions throughout your household. Place safety gates at both the upper and lower ends of staircases, use outlet covers for electrical sockets, and anchor bulky furniture to the wall to minimize the risk of tipping.
- **Secure Hazardous Items:** Keep dangerous substances, such as cleaning supplies and medications, out of reach. Use child-resistant locks on cabinets containing these items.
- **Window Safety:** Install window guards or stops to minimize the risk of falls. Keep furniture at a distance from windows to discourage climbing.
- **Education:** Educate children about potential hazards and establish clear rules for their safety. Teach them about emergency procedures, including how to call for help and escape routes.
- **Supervision:** Maintain vigilant supervision, especially around potentially dangerous areas like kitchens, bathrooms, and swimming pools. Never leave young children unattended, even for a short period.
- **Emergency Contacts:** Ensure that emergency contact information is easily accessible. Educate children on how to call emergency services and furnish them with a roster of contact numbers for reliable adults.

Elderly Safety

The elderly often face different challenges, including mobility issues and potential health concerns. Tailoring safety strategies to their needs is crucial for maintaining their well-being.

- **Fall Prevention:** Elderly individuals often face the risk of falls. Enhance safety by installing handrails along staircases, in bathrooms, and near beds. Employ non-slip mats in the bathroom and explore options like walk-in showers or baths.
- **Well-Lit Spaces:** Ensure that living spaces are well-lit to aid visibility. Install nightlights in hallways and bathrooms to prevent tripping during nighttime movements.
- **Medication Management:** Establish a reliable system for medication management. Use pill organizers and set up a schedule for administering medications. Consider installing reminder systems or alarms.
- **Emergency Response Plan:** Develop a clear emergency response plan. Ensure that the elderly family member has access to emergency numbers and knows the procedures to follow in case of a health emergency or natural disaster.
- **Accessible Living Spaces:** Modify living spaces to accommodate mobility challenges. Consider installing ramps for easy access, grab bars in bathrooms, and raised toilet seats for comfort.
- **Regular Health Checkups:** Schedule regular health checkups and screenings. Stay proactive about addressing any health concerns and adjusting safety measures accordingly.
- **Social Engagement:** Encourage social engagement to prevent isolation. Regular interaction with friends, family, or community groups contributes to emotional well-being and can provide additional support.

Personalized Safety Measures

Recognizing the individual needs and preferences of vulnerable family members allows for further customization of safety measures.

- **Communication Tools:** Provide communication tools that suit the capabilities of each individual. This may include simplified phones for the elderly or establishing alternative communication methods for children, such as walkie-talkies.
- **Personal Alarms:** Consider personal alarms for both children and the elderly. These devices can be discreetly worn and used to alert others in case of an emergency or if assistance is needed.
- **Identifying Safe Spaces:** Teach both children and the elderly about designated safe spaces within the home. These are areas where they can go during emergencies or if they feel threatened.
- **Establishing Routine Check-Ins:** Implement a routine check-in system, especially for elderly family members living alone. Regular phone calls, video chats, or visits provide reassurance and an opportunity to address any emerging concerns.

- **Technology for Monitoring:** Explore technological solutions for monitoring vulnerable family members. Smart home devices, wearables, or security cameras can provide additional layers of safety and peace of mind.

Fire Safety Measures

Fires pose a significant risk to households, and customized strategies are crucial for the safety of children and the elderly.

- **Fire Drills:** Conduct regular fire drills with both children and the elderly. Practice evacuation routes, use of fire extinguishers, and meeting points outside the home.
- **Fire Safety Education:** Educate family members about fire safety. Teach children how to stop, drop, and roll. Ensure that the elderly are aware of fire hazards and the proper use of fire safety equipment.
- **Accessible Fire Extinguishers:** Place easily accessible fire extinguishers in key areas of the home. Ensure that family members know how to use them and conduct regular checks to confirm their functionality.
- **Emergency Escape Ladders:** If living in a multi-story home, consider emergency escape ladders for upper floors. Practice using them during fire drills.
- **Smoke and Carbon Monoxide Detectors:** Ensure the safety of your home by placing smoke and carbon monoxide detectors in key areas. Be diligent in inspecting and changing batteries regularly to maintain peak performance.

Weather and Natural Disaster Preparedness

Customized safety strategies should also address the specific needs of vulnerable family members during weather-related emergencies or natural disasters.

- **Emergency Kits:** Prepare personalized emergency kits for both children and the elderly. Include essential items such as medications, comfort items, and necessary supplies.
- **Evacuation Plans:** Develop clear evacuation plans tailored to the capabilities of each family member. Consider accessibility requirements and coordinate with local authorities for additional support.
- **Communication Plans:** Establish communication plans during emergencies. Teach children how to use emergency communication tools, and ensure that the elderly have a reliable means of staying connected.
- **Safe Areas in the Home:** Identify safe areas within the home for different types of emergencies. These spaces should be easily accessible and equipped with emergency supplies.

- **Community Support Networks:** Foster relationships within the community to provide additional support during emergencies. Connect with neighbors and local organizations to establish a support network.

Haven't done it yet?

Scan the QR code and get direct access to 80 amazing self-sufficiency projects! Start now and see instant results—don't miss out on the opportunity to become more independent today!

Chapter 6:

Role of Guard Dogs

The role of guard dogs in home security goes beyond their reputation as loyal companions. When properly trained, dogs can serve as effective deterrents, alert systems, and, if needed, as protectors. Understanding the nuances of training and utilizing guard dogs can significantly enhance your home security. Let's delve into the practical aspects of the role of guard dogs, including training methods and their contributions to a secure home environment.

1. **Natural Instincts and Breeds**

Certain dog breeds possess inherent qualities that make them well-suited for guarding and protecting. Breeds like German Shepherds, Doberman Pinschers, Rottweilers, and Belgian Malinois are known for their intelligence, loyalty, and protective instincts. However, it's crucial to consider the individual temperament of each dog, as not all members of a breed may exhibit the same traits.

2. **Training for Guarding**

Proper training is essential to harness a dog's natural protective instincts and turn them into effective guard dogs. Key elements of guard dog training include:

- **Obedience Training:** Start with basic obedience training to establish a foundation of discipline and responsiveness. Commands like sit, stay, and come are fundamental.
- **Socialization:** Introduce the dog to various environments, people, and situations from an early age. Socialization helps prevent aggressive behavior towards non-threatening individuals and enhances the dog's adaptability.
- **Protection Training:** Gradually introduce protection training to develop the dog's guarding skills. This involves teaching the dog to alert, deter, and protect in response to perceived threats.
- **Bite Inhibition:** Train the dog in bite inhibition to ensure controlled and measured responses. A well-trained guard dog should know when and how much force is appropriate.
- **Alertness and Distraction Training:** Teach the dog to differentiate between normal activity and potential threats. Use controlled scenarios to expose the dog to various stimuli and reinforce appropriate responses.

3. **Deterrence and Alert System**

Guard dogs serve as a deterrent simply by their presence. Their protective instincts and watchful nature create an added layer of security, dissuading potential intruders. Additionally, they function as efficient alert systems:

- **Barking Alerts:** Dogs are naturally inclined to bark when they sense something unusual. This barking serves as an audible alert, notifying homeowners of potential threats or disturbances.
- **Heightened Senses:** Dogs have keen senses of smell and hearing. Their ability to detect changes in the environment, such as unfamiliar scents or noises, contributes to an early warning system.
- **Territorial Instincts:** Guard dogs develop a sense of territory and ownership over their home environment. This territorial instinct makes them more vigilant and responsive to perceived threats within their domain.

4. **Integration with Home Security Systems**

Guard dogs can complement modern home security systems to create a comprehensive protective network. Integrating guard dogs with technological solutions enhances the overall effectiveness of your security measures:

- **Surveillance Cameras:** Combine guard dog presence with strategically placed surveillance cameras. This provides visual confirmation of potential threats and aids in assessing the situation.
- **Motion Sensors:** Use motion sensors in key areas to detect movement. When paired with guard dogs, these sensors can trigger alerts and provide real-time information about the source of activity.
- **Smart Home Integration:** Leverage smart home technology to monitor your property remotely. This allows you to check in on your guard dog, receive alerts, and assess the security of your home from anywhere.

5. **Responsibilities and Care**

Owning a guard dog comes with certain responsibilities and care considerations:

- **Health Maintenance:** Regular veterinary check-ups, vaccinations, and a balanced diet contribute to the overall health and well-being of the guard dog.
- **Adequate Exercise:** Guard dogs, particularly those with high energy levels, require regular exercise. Physical activity not only maintains their health but also helps channel their energy in positive ways.
- **Bonding and Affection:** Establishing a strong bond with the guard dog is crucial. Dogs thrive on companionship, attention, and affection. A well-bonded dog is more likely to be loyal and responsive.
- **Secure Containment:** Ensure that the dog is securely contained within the property. This may involve proper fencing, designated areas, or other containment measures.
- **Proper Training Reinforcement:** Regular reinforcement of training is essential. Periodic training sessions help maintain the dog's skills and responsiveness.

6. **Legal Considerations**

Understanding the legal aspects of owning a guard dog is important:

- **Liability:** Homeowners are generally responsible for the actions of their dogs. It's crucial to be aware of local laws and regulations regarding dog ownership and liability.
- **Proper Signage:** Displaying signs indicating the presence of a guard dog can serve as a precautionary measure. This informs visitors and potential intruders about the security measures in place.

7. **Consideration of Family Dynamics**

When integrating a guard dog into a family environment, it's essential to consider family dynamics, especially if there are children or elderly family members:

- **Child-Friendly Training:** Ensure that the guard dog is trained to be friendly and gentle with children. Supervise interactions and teach children how to interact appropriately with the dog.
- **Compatibility with Other Pets:** Consider the compatibility of a guard dog with other pets in the household. Proper introductions and gradual acclimatization can help prevent conflicts.
- **Sensitivity to Elderly Members:** Be mindful of the needs and sensitivities of elderly family members. If the dog is particularly large or energetic, take precautions to prevent unintentional knocks or falls.

Chapter 7:

Communication Protocols

Effective crisis communication is essential in navigating challenging situations and ensuring that information is conveyed clearly, timely, and accurately. Whether facing a natural disaster, a medical emergency, or any crisis scenario, having well-established communication protocols is crucial for minimizing confusion, maintaining order, and ensuring the safety and well-being of individuals involved. Let's explore practical methods and strategies for effective crisis communication.

1. **Clear Chain of Command**

Establishing a clear chain of command is foundational for effective crisis communication. Define roles and responsibilities within the organization or community, designating individuals or teams responsible for communication at different levels. This ensures that information flows in a structured manner, preventing delays and reducing the risk of miscommunication.

2. **Rapid Information Gathering**

In a crisis, timely information is paramount. Implement systems for rapid information gathering from reliable sources. This may involve coordinating with emergency services, monitoring official channels, and utilizing technology to collect real-time data. The quicker and more accurately information is gathered, the better the response can be tailored to the situation.

3. **Centralized Communication Hub**

Designate a centralized communication hub where information is collated, verified, and disseminated. This hub serves as the primary point for coordinating communication efforts, allowing for consistency in messaging and preventing the spread of misinformation. Utilize technology to facilitate swift communication within the hub.

4. **Multichannel Communication**

Adopt a multichannel communication approach to reach a diverse audience. Utilize a combination of communication channels such as:

- **Traditional Media:** Broadcast information through radio, television, and newspapers to reach a broad audience.
- **Social Media:** Leverage social media platforms for real-time updates and interaction. Platforms like Twitter, Facebook, and Instagram are effective for reaching a wide audience quickly.
- **Text Messages and Alerts:** Implement mass text messaging or alert systems to notify individuals directly. This is especially useful for targeted communication to specific groups or communities.
- **Public Announcements:** Utilize public address systems, loudspeakers, or sirens for on-the-ground communication in localized areas.

5. **Consistent Messaging**

Consistency in messaging is crucial to avoid confusion and build trust. Ensure that information provided through different channels and by various spokespersons aligns seamlessly. Consistent messaging helps establish credibility and prevents the spread of conflicting information.

6. **Tailored Communication for Audiences**

Recognize the diversity of your audience and tailor communication accordingly. Consider language preferences, cultural sensitivities, and accessibility needs. Providing information in multiple languages and formats ensures that the message is understood by a wide range of individuals, including those with disabilities.

7. **Two-Way Communication**

Promote two-way communication channels to facilitate the exchange of information. Encourage feedback, questions, and concerns from the community or affected individuals. Establish hotlines, online forums, or community meetings where people can voice their thoughts and receive clarifications.

8. **Training and Drills**

Regularly conduct training sessions and crisis communication drills. This prepares communication teams and individuals to respond effectively under pressure. Simulating crisis scenarios allows for the identification of weaknesses in communication protocols and provides an opportunity for improvement.

9. **Transparency and Honesty**

Maintain a commitment to transparency and honesty. In crisis situations, individuals rely on accurate information to make informed decisions. Concealing or manipulating information erodes trust and can lead to increased panic. Communicate openly about the situation, acknowledging uncertainties while emphasizing the steps being taken to address the crisis.

10. **Crisis Communication Plan**

Create an all-encompassing plan for crisis communication that details the roles, duties, and communication protocols that will be followed. This plan should include:

- **Predefined Messages:** Draft templates for different types of crises to facilitate quick and consistent communication.
- **Contact Lists:** Maintain updated contact lists for internal and external stakeholders, including emergency services, media contacts, and community leaders.
- **Social Media Protocols:** Establish guidelines for social media usage during a crisis, including approved hashtags, official accounts, and measures to counteract misinformation.
- **Media Relations:** Outline procedures for engaging with the media, including designated spokespersons, press releases, and press conferences.

11. **Use of Technology**

Leverage technology to enhance communication capabilities:

- **Emergency Alert Systems:** Implement automated emergency alert systems that can deliver messages via text, phone calls, or sirens to a wide audience.
- **GIS (Geographic Information System):** Use GIS technology to map and visualize crisis situations, helping responders and the public understand the geographic impact of the event.
- **Mobile Apps:** Develop or utilize mobile applications that provide real-time updates, emergency resources, and communication features.

12. **Post-Crisis Communication**

Effective communication extends beyond the crisis itself. Develop a post-crisis communication strategy to:

- **Provide Updates:** Keep the community informed about recovery efforts, ongoing support, and any long-term impacts.
- **Address Concerns:** Acknowledge and address any concerns or grievances that may have arisen during the crisis. Transparency in the recovery phase is as crucial as during the crisis.

- **Share Lessons Learned:** Conduct a thorough review of the crisis response, identify areas for improvement, and share lessons learned with stakeholders. This contributes to continuous improvement in communication protocols.

13. Community Involvement

Involve the community in the communication process. Collaborate with local leaders, community organizations, and residents to ensure that communication strategies are culturally sensitive, relevant, and inclusive.

14. Emotional Support Communication

Acknowledge the emotional impact of a crisis and incorporate empathy into communication. Provide information about available support services, counseling resources, and community assistance. Addressing emotional needs fosters resilience and community cohesion.

Chapter 8:

Outside Excursions

L eaving the safety of your home during a crisis requires careful planning and consideration of potential risks. Whether you're evacuating due to a natural disaster, facing a sudden emergency, or navigating challenging circumstances, adhering to safety tips for outside excursions is crucial for protecting yourself and your loved ones. Let's explore practical advice to ensure a secure and well-prepared departure when leaving home during crises.

1. **Emergency Kit Essentials**

Before venturing outside during a crisis, ensure you have a well-equipped emergency kit. This kit should include:

- **First Aid Supplies:** Basic medical supplies, prescription medications, and a first aid manual.
- **Water and Non-Perishable Food:** Carry sufficient water for each person, as well as non-perishable snacks or meal replacement bars.
- **Important Documents:** Store crucial documents like identification, insurance policies, and medical records in a water-resistant container as a precaution.
- **Flashlights and Batteries:** Pack reliable flashlights with extra batteries to ensure visibility, especially in low-light conditions.
- **Clothing and Personal Items:** Include weather-appropriate clothing, sturdy footwear, and personal hygiene items.
- **Communication Devices:** Bring fully charged cell phones, portable chargers, and, if possible, a battery-powered or hand-crank radio for updates.

2. **Stay Informed**

Stay informed about the crisis and follow official updates from relevant authorities. Listen to emergency broadcasts on the radio, check official websites or social media channels, and adhere to evacuation orders or advisories issued by local authorities.

3. Evacuation Routes

Familiarize yourself with evacuation routes well in advance. Plan multiple routes to account for unexpected road closures or congestion. Use navigation apps or GPS devices to help guide you if needed.

4. Inform Others

Let someone know about your plans and intended destination. Share your itinerary, including planned stops or meeting points, with a friend, family member, or neighbor. Regularly communicate your whereabouts, especially if plans change.

5. Travel in Groups, if Possible

Whenever feasible, travel in groups. There is safety in numbers, and having others with you can provide support, assistance, and additional resources. Stick together and coordinate your movements during outside excursions.

6. Obey Authorities' Instructions

Follow the instructions given by emergency responders, law enforcement, and other authorities. If evacuation orders are issued, comply promptly and follow designated routes. Authorities have access to real-time information and can guide you away from potential dangers.

7. Be Aware of Your Surroundings

Maintain situational awareness at all times. Pay attention to your surroundings, stay vigilant, and be mindful of potential hazards. Avoid areas prone to flooding, landslides, or other risks associated with the crisis.

8. Travel Light

Keep your load manageable by traveling light. Only bring essential items from your emergency kit, focusing on necessities that will sustain you during the excursion. A lighter load makes it easier to move quickly and efficiently.

9. Dress Appropriately

Wear weather-appropriate clothing and sturdy footwear. Dress in layers to adapt to changing conditions, and consider packing a rain poncho, hat, or other protective gear based on the weather forecast.

10. Use Caution with Vehicles

If using a vehicle during an evacuation, exercise caution and adhere to traffic rules. Plan for potential fuel shortages, heavy traffic, or road closures. Ensure your vehicle is in good condition and has sufficient fuel.

11. Plan for Pet Safety

If you have pets, include them in your evacuation plans. Bring pet supplies, such as food, water, leashes, and carriers. Identify pet-friendly shelters or accommodations in advance.

12. Carry Cash

Have a small amount of cash on hand. In crisis situations, electronic payment systems may be unreliable, and having cash can be useful for purchasing necessities or covering unexpected expenses.

13. Monitor Health and Well-Being

Prioritize your health and well-being during outside excursions. Stay hydrated, take breaks if needed, and be mindful of any physical or emotional stress. Attend to any medical needs promptly.

14. Secure Your Home

Before leaving, secure your home to the best of your ability. Close and lock doors and windows, turn off utilities if instructed, and unplug electronic devices to prevent electrical fires.

15. Have a Reunification Plan

Establish a reunification plan with family or loved ones in case you get separated during the evacuation. Determine meeting points and communication methods to ensure everyone reconnects safely.

16. Adapt to Changing Circumstances

Be flexible and adaptable in your plans. Crisis situations can evolve rapidly, and conditions may change unexpectedly. Stay informed and be prepared to adjust your course of action based on the latest information.

17. Maintain Calm and Composure

During outside excursions, maintaining a calm and composed demeanor is essential. Panic can exacerbate the situation and lead to poor decision-making. Stay focused, follow your plan, and encourage those around you to remain calm.

18. Stay Connected

Keep your communication devices charged and stay connected with others in your group. Regularly check in with each other to ensure everyone is accounted for and informed.

19. Be a Good Samaritan

Help others in need if it is safe to do so. Extend assistance to those who may require support, especially vulnerable individuals such as the elderly, children, or those with mobility challenges.

20. Follow Public Health Guidelines

In situations where health crises are involved, such as pandemics, adhere to public health guidelines. Practice social distancing, wear masks if recommended, and follow hygiene practices to protect yourself and others.

BOOK 4:
SURVIVAL SHELTER GUIDE

Chapter 1:

Shelter Construction

Survival shelter construction is a fundamental skill that can make a significant difference in your ability to endure challenging situations. Whether you're facing the wilderness, natural disasters, or unexpected emergencies, knowing how to build different types of shelters is crucial.

Lean-To Shelter

Materials Needed:

- Long, sturdy branch or pole
- Smaller branches or sticks
- Leaves, grass, or foliage

Construction Steps:

1. Find a long, sturdy branch or pole and secure one end to a stable anchor like a tree trunk or a rock.

2. Lean the other end of the pole against a solid support, creating an inclined angle.

3. Place smaller branches or sticks against the main pole, creating a framework for the shelter.

4. Cover the framework with leaves, grass, or foliage to provide insulation and protection from the elements.

5. Ensure the shelter is wide enough to accommodate your body while lying down.

Debris Hut Shelter

Materials Needed:

- Long, sturdy branches
- Leaves, grass, or foliage
- Vines or cordage

Construction Steps:

1. Create a A-frame structure using long, sturdy branches, leaning them against each other at the top.

2. Secure the frame by tying the branches together at the top using vines or cordage.

3. Add additional branches to create a rib-like structure along the sides of the A-frame.

4. Cover the frame with leaves, grass, or foliage, creating a thick layer for insulation.

5. Seal any gaps in the debris hut to minimize heat loss.

A-Frame Shelter

Materials Needed:

- Long, sturdy branches
- Cordage or vines
- Tarp or poncho (optional)

Construction Steps:

1. Arrange two long, sturdy branches in the shape of an "A," crossing at the top.

2. Secure the crossed point with cordage or vines, ensuring stability.

3. Place additional branches along both sides of the "A" to form a triangular structure.

4. Optionally, drape a tarp or poncho over the frame for additional protection.

5. Secure the covering with more cordage or by weighing down the edges with rocks.

Snow Cave Shelter

Materials Needed:

- Snow shovel or improvised digging tools

Construction Steps:

1. Find a snowdrift or compacted snow.

2. Dig into the snow to create an entrance, ensuring it is below the sleeping platform to trap warm air.

3. Dig a horizontal tunnel into the snow to form a sleeping platform, leaving a thickness of at least 12 inches for insulation.

4. Create a ventilation hole at the top of the snow cave to allow fresh air to circulate.

5. Smooth the interior surfaces to reduce dripping from melting snow.

Tarp Shelter

Materials Needed:

- Tarp or poncho
- Cordage or paracord
- Stakes or heavy objects

Construction Steps:

1. Lay the tarp or poncho flat on the ground.

2. If using a rectangular tarp, fold it diagonally to create a triangular shape.

3. Use cordage or paracord to tie the top corners of the triangular tarp to nearby trees or structures.

4. Secure the bottom corners of the tarp to the ground using stakes or heavy objects.

5. Adjust the height and tension to create a shelter with a comfortable living space.

Teepee Shelter

Materials Needed:

- Long, sturdy poles
- Cordage or vines
- Leaves, grass, or foliage

Construction Steps:

1. Gather long, sturdy poles and tie them together at the top using cordage or vines.

2. Spread the poles out, forming a conical shape.

3. Add additional poles around the structure to create a framework.

4. Cover the framework with leaves, grass, or foliage to provide insulation and camouflage.

5. Leave an opening at the top for ventilation and a small entrance at the bottom.

Bivy Bag Shelter

Materials Needed:

- Bivy bag or survival sleeping bag
- Ground tarp or mat

Construction Steps:

1. Choose a level and sheltered area.

2. Lay down a ground tarp or mat to insulate against the cold ground.

3. Enter the bivy bag and zip it up, ensuring your entire body is covered.

4. Use the hood of the bivy bag to cover your head and protect against the elements.

5. Adjust the bag's ventilation openings to regulate temperature and humidity.

Chapter 2:

Shelter Hardening

Shelter hardening involves reinforcing and fortifying your shelter to enhance its durability and security. Whether you are in a wilderness survival scenario or facing the aftermath of a disaster, ensuring that your shelter can withstand environmental challenges and potential threats is crucial for your safety and well-being. Here, we'll explore practical strategies for shelter hardening, focusing on strengthening your shelter's structure and improving its ability to protect you in adverse conditions.

Structural Reinforcement

- **Additional Support:** Strengthen the framework of your shelter by adding extra support to key areas. This can involve securing additional branches, logs, or even improvised materials to reinforce the existing structure. Pay special attention to corners and joints where extra support is often needed.
- **Cross-Bracing:** Implement cross-bracing by securing diagonal supports across the structure. This technique helps distribute weight and forces more evenly, reducing the risk of collapse during heavy winds or storms.
- **Lashings and Bindings:** Use durable cordage or vines to create tight lashings and bindings. Secure all components of your shelter firmly to prevent shifting or separation during adverse weather conditions. Properly tied lashings add structural integrity to your shelter.

Weatherproofing

- **Waterproofing:** Apply waterproofing agents to materials covering your shelter, such as tarps, ponchos, or natural coverings like leaves and foliage. Waterproofing helps prevent leaks and ensures that your shelter remains dry during rain or snow.
- **Elevated Flooring:** If feasible, create an elevated platform or floor within your shelter. This prevents moisture from seeping through the ground and provides a more comfortable and dry sleeping area. Elevating your sleeping space can be especially beneficial in damp or flood-prone environments.
- **Windproofing:** Position your shelter to minimize its exposure to prevailing winds. This simple adjustment can significantly reduce the impact of strong gusts on your shelter.

Additionally, use windbreaks such as natural barriers or constructed walls to shield your shelter.

Security Measures

- **Camouflage:** Camouflage your shelter to blend in with the surrounding environment. Use natural materials, such as branches, leaves, and mud, to disguise your shelter and make it less noticeable to potential threats.
- **Concealed Entrances:** Design your shelter with a concealed entrance to minimize visibility. This adds an element of surprise and security, making it more challenging for intruders or wildlife to detect your presence.
- **Warning Systems:** Set up warning systems around your shelter, such as noise-making devices or strategically placed objects. These systems can alert you to approaching threats, giving you valuable time to assess the situation and take appropriate action.

Fire Safety

- **Fire-Resistant Materials:** Whenever possible, use fire-resistant materials in your shelter construction. Certain types of rocks, treated fabrics, or fire-resistant coatings can reduce the risk of accidental fires within or near your shelter.
- **Safe Fireplace:** If you plan to have a fire inside your shelter, create a safe fireplace area. Use fire-resistant stones or construct a makeshift fireplace with proper ventilation to reduce the risk of fire spreading.

Wildlife Deterrence

- **Food Storage:** Keep food securely stored and away from your shelter to deter wildlife. Hanging food in a tree or using bear-resistant containers can prevent animals from being attracted to your shelter.
- **Noise and Movement:** Avoid making unnecessary noise and sudden movements around your shelter. This helps minimize the chances of attracting wildlife, especially during the night when many animals are active.

Emergency Exits

- **Multiple Exits:** Design your shelter with multiple exits for emergency situations. This ensures that you can quickly evacuate if needed, especially if your primary entrance is compromised or blocked.
- **Clear Escape Paths:** Keep escape paths clear of obstacles to facilitate a swift exit. Regularly check the condition of these paths to ensure they remain viable in case of an emergency.

Long-Term Considerations

- **Sustainable Materials:** If you anticipate a prolonged stay in your shelter, consider using sustainable and durable materials for construction. This might include using treated or naturally resistant wood, as well as materials that withstand prolonged exposure to the elements.
- **Regular Maintenance:** Perform regular maintenance on your shelter to address wear and tear. Inspect lashings, bindings, and structural components to identify any weaknesses. Timely repairs and upkeep contribute to the long-term durability of your shelter.

Strategic Lighting

- **Low-Profile Lighting:** Use low-profile and diffuse lighting sources inside your shelter. This minimizes the visibility of light from a distance, reducing the likelihood of attracting unwanted attention during the night.
- **Light Discipline:** Practice light discipline by limiting the use of artificial lighting. If possible, use natural light during the day and rely on minimal lighting sources at night to maintain a low profile.

Elevated Observation Points

- **Lookout Platforms:** If your shelter is in a location that allows for it, create an elevated lookout platform. This offers a vantage point for observing the surrounding area, providing early detection of potential threats.
- **Secure Perimeter:** Establish a secure perimeter around your shelter with clear lines of sight. Remove obstacles and potential hiding spots for intruders, improving overall security.

Chapter 3:

Wild Shelters

Wild shelters, crafted from the materials nature provides, are essential skills for anyone venturing into the great outdoors or finding themselves in a survival situation. Learning techniques for makeshift shelters using natural resources allows individuals to adapt to their surroundings and create a protective haven.

Debris Hut

The debris hut is a classic and effective shelter made from natural materials. To construct a debris hut:

1. Find a sturdy, long branch and use it as the ridgepole.

2. Lean one end of the ridgepole against a sturdy base, such as a rock or tree.

3. Create a rib-like framework along the sides of the ridgepole using smaller branches.

4. Cover the framework with leaves, grass, or any available debris for insulation.

5. Ensure the entrance is small to retain heat and protect against wind and rain.

Lean-To Shelter

A lean-to shelter is simple to build and provides protection from the wind and rain. To construct a lean-to:

1. Find a long, sturdy branch and secure one end to a stable anchor, such as a tree or rock.

2. Angle the branch to the ground, creating an inclined roof.

3. Place smaller branches or sticks against the main pole, forming a framework.

4. Cover the framework with leaves, grass, or debris for insulation.

5. Ensure the open side of the lean-to faces away from prevailing winds.

A-Frame Shelter

The A-frame shelter is easy to construct and offers stability. To build an A-frame shelter:

1. Use two long, sturdy branches to create an "A" shape by crossing them at the top.

2. Secure the crossed point with vines or cordage to ensure stability.

3. Add additional branches along both sides of the "A" to form a triangular structure.

4. Cover the framework with leaves, grass, or debris for insulation.

5. Adjust the height and width based on personal preferences and environmental conditions.

Tarp or Poncho Shelter

Utilizing a tarp or poncho is a quick way to create a makeshift shelter. To set up a tarp or poncho shelter:

1. Find a suitable location and attach one corner of the tarp or poncho to a stable anchor, such as a tree.

2. Use cordage to secure the opposite corner to another anchor, creating a ridgeline.

3. Pull the sides taut and secure them to the ground using stakes or heavy objects.

4. Adjust the height and tension to create a shelter with a comfortable living space.

Bushcraft Shelter

Bushcraft shelters involve using natural materials like branches, leaves, and debris. To create a bushcraft shelter:

1. Gather long, flexible branches and weave them into a framework or lattice.

2. Secure the framework in place using vines, cordage, or natural fibers.

3. Cover the framework with leaves, grass, or debris for insulation.

4. Optimize the design based on the available materials and the specific environment.

Snow Cave Shelter

In snowy environments, a snow cave offers insulation and protection. To construct a snow cave shelter:

1. Find a snowdrift or compacted snow.

2. Dig into the snow to create an entrance, ensuring it is below the sleeping platform to trap warm air.

3. Dig a horizontal tunnel into the snow to form a sleeping platform.

4. Create a ventilation hole at the top to allow fresh air circulation.

5. Smooth the interior surfaces to prevent dripping from melting snow.

Raised Bed Shelter

Building a raised bed provides insulation from the cold ground. To create a raised bed shelter:

1. Construct a frame using branches, logs, or rocks, elevating it from the ground.

2. Fill the frame with insulating materials such as leaves, grass, or pine needles.

3. Create a sleeping platform on top of the raised bed.

4. Cover the bed with additional insulating materials for comfort and warmth.

Rock Shelter

In rocky terrains, natural rock formations can be used to create a shelter. To utilize a rock shelter:

1. Find a sheltered area beneath an overhang or between large rocks.

2. Use smaller rocks to create a windbreak or partial walls.

3. Create a raised bed or insulating layer using available materials.

4. Ensure the chosen location provides protection from the elements.

Tree Well Shelter

Tree wells, the depressed areas around the base of trees, can be utilized for shelter. To create a tree well shelter:

1. Identify a tree with a well-defined tree well.

2. Use branches, leaves, and debris to create a bedding area within the well.

3. Arrange additional materials to form a protective barrier around the well.

4. Utilize the tree's canopy for added overhead protection.

Cave Shelter

If available, caves offer natural protection. To use a cave as a shelter:

1. Locate a cave with a safe and accessible entrance.

2. Clear any debris or potential hazards inside the cave.

3. Create a sleeping area away from the entrance to avoid drafts.

4. Assess the cave for signs of wildlife or potential dangers.

Considerations for Wild Shelters:

- **Location Matters:** Choose a shelter location that provides natural protection from the elements and minimizes exposure to wind and rain. Assess the ground for potential water runoff and flooding.
- **Ventilation:** Ensure proper ventilation in your shelter to prevent condensation and maintain air circulation. This is crucial for avoiding moisture build-up and mold.
- **Fire Safety:** If you plan to have a fire near your shelter, consider the proximity to flammable materials and the potential for sparks. Establish a safe area for a controlled fire if needed.
- **Adaptability:** Be adaptable in your shelter-building approach based on available resources, weather conditions, and the specific environment. Use what nature provides to your advantage.

- **Wildlife Awareness:** Be mindful of wildlife and their habitats. Choose shelter locations that minimize disturbance to local fauna, and take precautions to avoid encounters with potentially dangerous animals.
- **Leave No Trace:** Adhere to Leave No Trace principles by minimizing your environmental impact. Upon departing from your shelter, disassemble it and return the area to its natural state.

Chapter 4:

Anti-atomic Shelters

The prospect of nuclear events has led to increased interest in anti-atomic shelters, commonly known as nuclear bunkers. These structures are designed to provide protection against the devastating effects of nuclear explosions, offering a haven for individuals and communities to seek refuge in times of crisis. Understanding the principles behind anti-atomic shelters and learning how to build a nuclear bunker is crucial for those looking to enhance their preparedness in the face of potential nuclear threats.

Before delving into the construction of anti-atomic shelters, it's essential to understand the nature of nuclear threats and the potential consequences of nuclear detonations. Nuclear explosions release vast amounts of energy, leading to immediate dangers such as blast waves, thermal radiation, and ionizing radiation. Fallout, composed of radioactive particles, poses a longer-term hazard, necessitating the need for effective sheltering strategies.

Key Components of Anti-Atomic Shelters

1. **Blast Protection:** Anti-atomic shelters must be structurally reinforced to withstand the intense pressure generated by nuclear blasts. This involves reinforced concrete walls and ceilings designed to absorb and deflect the force of the blast.

2. **Radiation Shielding:** Shielding against ionizing radiation is critical for a nuclear bunker. Materials with high-density properties, such as lead or concrete, are effective in absorbing and blocking radiation. The thicker the shielding, the greater the protection.

3. **Ventilation Systems:** To ensure a continuous supply of fresh air while preventing the entry of radioactive particles, anti-atomic shelters incorporate ventilation systems equipped with high-efficiency particulate air (HEPA) filters. These filters trap airborne radioactive particles.

4. **Water and Food Storage:** Nuclear bunkers should include provisions for storing ample water and non-perishable food supplies. In the aftermath of a nuclear event, access to uncontaminated water and food becomes essential for survival.

5. **Sanitation Facilities:** Adequate sanitation facilities, such as toilets and waste disposal systems, are vital for maintaining hygiene within the bunker. Effective waste management is essential to mitigate the risk of disease transmission in enclosed environments.

6. **Communication Systems:** Anti-atomic shelters should be equipped with communication systems that allow occupants to stay informed about the external environment. Radio receivers capable of receiving emergency broadcasts are essential.

7. **Power Generation:** A reliable power source is necessary for lighting, ventilation, and electronic equipment. Backup power generators, as well as alternative energy sources like solar power, contribute to the sustainability of the bunker.

Building an Anti-Atomic Shelter

Constructing a nuclear bunker requires careful planning and adherence to specific construction principles. While the specifics may vary based on individual needs and available resources, the following general steps provide a foundation for building an effective anti-atomic shelter:

1. **Site Selection:** Choose a location for the bunker that minimizes exposure to potential targets and fallout. Underground sites, such as basements or specially constructed bunkers, offer enhanced protection.

2. **Reinforced Construction:** Build the shelter with reinforced materials, focusing on the structural integrity of walls and ceilings. Reinforced concrete is commonly used due to its ability to withstand blast pressure.

3. **Shielding Design:** Incorporate shielding materials into the construction to protect against radiation. Calculate the required thickness based on the type and intensity of radiation expected in the area.

4. **Ventilation System Installation:** Install a ventilation system with HEPA filters to ensure a continuous supply of clean air. The system should be capable of sealing off inlets during periods of high radiation to prevent contamination.

5. **Water and Food Storage:** Allocate dedicated spaces for water and food storage. Use containers that can seal tightly to prevent contamination. Rotate food supplies regularly to maintain freshness.

6. **Sanitation Facilities:** Design and install sanitation facilities, including toilets and waste disposal systems. Incorporate measures for waste treatment and containment.

7. **Communication Systems Setup:** Install communication systems that allow occupants to receive emergency broadcasts and communicate with the outside world if necessary. Ensure redundancy in communication equipment.

8. **Power Generation Solutions:** Integrate power generation solutions, such as backup generators and renewable energy sources, to sustain the bunker's power needs. Batteries and energy storage systems provide additional support.

9. **Security Measures:** Implement security measures to control access to the bunker. Consider reinforced entry points and surveillance systems to enhance overall safety.

10. **Emergency Protocols and Training:** Establish clear emergency protocols for bunker occupants. Conduct regular drills and training sessions to familiarize individuals with the shelter's layout and emergency procedures.

Maintaining and Updating the Bunker

Regular maintenance and updates are crucial for ensuring the ongoing effectiveness of an anti-atomic shelter. This involves:

- Regularly testing and replacing ventilation filters to maintain air quality.
- Monitoring radiation levels inside and outside the bunker.
- Conducting periodic structural inspections to identify and address any weaknesses.
- Updating communication systems to stay compatible with emerging technologies.
- Rotating and replenishing water and food supplies to ensure freshness.

Chapter 5:

DIY Bunkers

Constructing a personal bunker, also known as a Do-It-Yourself (DIY) bunker, requires careful planning, resourcefulness, and adherence to safety guidelines. A personal bunker can serve as a retreat during emergencies, providing protection against various threats such as natural disasters, civil unrest, or other unforeseen circumstances.

1. **Site Selection**

Selecting an appropriate location for your bunker is the first crucial step. Consider the following factors:

- **Geographical Location:** Choose an area away from potential threats, such as industrial facilities, military installations, or densely populated urban areas.
- **Soil Composition:** Opt for stable soil that facilitates excavation and provides structural support. Sandy or loamy soils may be easier to work with than rocky or clayey soils.

- **Water Table:** Ensure the chosen site has a low water table to minimize the risk of flooding. Excessive water levels pose a risk to the structural stability of your bunker.
- **Accessibility:** Select a site that is easily accessible, both for construction materials and for the eventual use of the bunker.

2. Legal Considerations

Before starting construction, research and comply with local building codes, zoning regulations, and any legal restrictions related to bunker construction. Secure the required permits to guarantee that your project complies with safety and legal regulations.

3. Design and Planning

Create a detailed design plan for your bunker, taking into account the following elements:

- **Size:** Determine the size of the bunker based on the number of occupants and intended use. Factor in enough space for storage, sleeping quarters, and necessary facilities.
- **Layout:** Plan the internal layout, including the placement of rooms, ventilation systems, emergency exits, and storage areas.
- **Structural Features:** Decide on the structural components, such as reinforced concrete walls and ceilings, to provide blast and radiation protection.

4. Excavation

Excavation is a critical phase that involves digging the bunker space into the ground. Follow these steps:

- **Mark the Area:** Use stakes and string to outline the dimensions of the bunker on the chosen site.
- **Excavate the Space:** Use heavy machinery, such as excavators, to dig the bunker space according to your design. Ensure the walls and ceiling are sloped for stability.
- **Debris Removal:** Regularly remove excavated soil and debris to maintain a clear workspace.

5. Reinforcement and Structure

Reinforce the bunker's structure to provide stability and protection:

- **Reinforced Concrete:** Pour and mold reinforced concrete to form the walls and ceiling. Reinforcement bars (rebar) can be added for additional strength.
- **Waterproofing:** In order to avoid water from penetrating the bunker walls, waterproofing materials should be applied to the exterior of the bunker walls.

6. Ventilation Systems

Install a ventilation system to ensure a continuous supply of fresh air and filter out contaminants:

- **Intake and Exhaust Vents:** Place vents strategically to allow fresh air intake and facilitate the expulsion of stale air.
- **HEPA Filters:** The removal of airborne particles and impurities can be accomplished through the installation of HEPA filters.

7. Entry and Exit Points

Design secure and accessible entry and exit points:

- **Bunker Door:** Install a sturdy and secure bunker door with multiple locking mechanisms. Consider reinforced steel doors for added security.
- **Emergency Exit:** Plan for an emergency exit, such as a secondary door or a hatch, for evacuation in case the primary entry is compromised.

8. Utilities and Systems

Incorporate essential utilities and systems for a functional bunker:

- **Power Supply:** Set up a reliable power supply using backup generators, solar panels, or alternative energy sources.
- **Water Storage:** Install water storage tanks or a well-designed water supply system to ensure a sustainable water source.
- **Sanitation Facilities:** Include sanitation facilities such as a toilet, waste disposal system, and washing area.
- **Emergency Lighting:** Install emergency lighting systems, including battery-powered or solar-powered lights, to ensure visibility during power outages.

9. Furnishing and Supplies

Furnish the bunker with necessary supplies and equipment:

- **Storage Areas:** Create designated storage areas for food, water, medical supplies, and other essential items.
- **Sleeping Quarters:** Set up sleeping quarters with comfortable bedding and insulation to maintain a comfortable temperature.
- **Communication Devices:** Equip the bunker with communication devices, such as a two-way radio or satellite phone, for emergency contact.

10. Security Measures

Implement security measures to protect the bunker and its occupants:

- **Surveillance Systems:** Install surveillance cameras and monitoring systems for added security.
- **Security Protocols:** Establish security protocols for access control and emergency situations.
- **Emergency Supplies:** Keep a stash of emergency supplies, including first aid kits, fire extinguishers, and tools.

11. **Testing and Maintenance**

Regularly test and maintain the bunker to ensure its functionality and readiness:

- **Ventilation Checks:** Test and replace ventilation filters regularly to maintain air quality.
- **Structural Inspections:** Periodically inspect the structural integrity of walls and ceilings.
- **Equipment Checks:** Verify the functionality of all systems, including power generators, water supplies, and communication devices.

If you're enjoying the journey so far, I'd be thrilled if you took a moment to share your thoughts with a honest review. Your feedback is not just valuable; it's what lights my path as an author. Thank you for being an essential part of this adventure!

BOOK 5:
SELF-DEFENSE AND FIRST AID CRASH COURSE

Chapter 1:

Fundamentals of Self-Defense

Understanding the fundamentals of self-defense is crucial for personal safety in various situations.

Basic Techniques

1. **Awareness and Prevention:** The foundation of self-defense lies in awareness. Understanding your surroundings and potential risks is the first line of defense. Avoiding potentially dangerous situations and being mindful of your environment can significantly reduce the likelihood of encountering threats.

2. **Situational Awareness:** Knowing what's happening around you at all times is key to personal safety. Practice observing your surroundings, identifying exits, and staying alert to any unusual behavior. Being proactive in assessing potential risks enhances your ability to respond effectively.

3. **Body Language and Confidence:** Projecting confidence through body language can deter potential attackers. Maintain an upright posture, make eye contact, and walk with purpose. Confidence can act as a deterrent, signaling to others that you are aware and prepared.

4. **Verbal De-escalation:** Effective communication is a powerful self-defense tool. Learning to defuse tense situations through calm and assertive verbal communication can prevent conflicts from escalating. Expressing boundaries and using clear, firm language can discourage potential threats.

5. **Escape and Evasion:** When faced with imminent danger, the ability to escape is crucial. Practice quick and decisive movements to create distance between yourself and a potential threat. Identify escape routes in unfamiliar environments, and be prepared to use them if necessary.

6. **Basic Strikes:** Understanding basic striking techniques provides a means of defending yourself when physical contact is unavoidable. Focus on simple and effective strikes, such as palm heel strikes, knee strikes, and elbow strikes. Practice these techniques to enhance muscle memory.

7. **Blocks and Parries:** Learning to block or parry incoming attacks is essential. Develop the skill to protect vital areas like the face, head, and torso. Use your arms and hands effectively to deflect and minimize the impact of strikes.

The Judicious Use of Weapons:

1. **Legal Considerations:** Before considering the use of weapons for self-defense, it's crucial to understand the legal implications. Familiarize yourself with local laws regarding the possession and use of self-defense tools, ensuring compliance with regulations.

2. **Pepper Spray:** Pepper spray is a widely used and effective non-lethal self-defense tool. It causes temporary incapacitation by irritating the eyes and respiratory system. Familiarize yourself with its proper use, aiming for the eyes of an assailant.

3. **Personal Alarms:** Personal alarms are valuable for drawing attention to a potential threat. Activating a loud alarm can deter attackers and alert nearby individuals to your situation. Keep the alarm easily accessible for quick deployment.

4. **Tactical Flashlights:** A powerful flashlight serves both as a tool for illuminating dark areas and a potential self-defense tool. Shine the bright light directly into an assailant's eyes to disorient them and create an opportunity for escape.

5. **Whistles:** Whistles serve as uncomplicated yet powerful instruments to capture attention. Employing a robust whistle can signal for assistance or discourage a potential threat. Carry a whistle on your keychain or in a readily accessible location.

6. **Personal Safety Apps:** Smartphone applications designed for personal safety can be valuable resources. These applications frequently encompass functionalities like emergency notifications, real-time location monitoring, and swift communication with authorities.

7. **Improvised Weapons:** Everyday items can serve as improvised weapons in self-defense situations. Objects like pens, keys, or umbrellas can be used strategically to create opportunities for escape or to incapacitate an attacker.

8. **Self-Defense Classes:** Enrolling in self-defense classes provides hands-on training in the use of various tools and techniques. Instructors can guide you in selecting and effectively using self-defense tools, ensuring that you are well-prepared in diverse scenarios.

Continued Training and Preparedness:

1. **Regular Practice:** Self-defense skills, like any other, require regular practice to maintain proficiency. Devote time to practicing techniques, drills, and scenarios to reinforce muscle memory and build confidence.

2. **Scenario-Based Training:** Simulate real-life scenarios during training to enhance your ability to respond effectively under stress. Scenario-based training helps bridge the gap between theory and practical application.

3. **Physical Fitness:** Maintaining physical fitness is integral to effective self-defense. Regular exercise enhances strength, agility, and overall endurance, contributing to your ability to evade threats and respond decisively.

4. **Mental Preparedness:** Mental resilience is as important as physical readiness. Develop a mindset that prioritizes personal safety, and cultivate the ability to stay calm and focused under pressure. Mental preparedness enhances decision-making in critical situations.

Chapter 2:

Creating a Survival Kit

Crafting a survival kit is a fundamental step toward preparedness for unforeseen circumstances. Whether facing natural disasters, outdoor adventures, or unexpected emergencies, a well-thought-out survival kit can be a lifesaver. The key is to include essential items that cover a range of scenarios, providing the tools needed to endure and overcome challenges.

Shelter and Warmth

In unpredictable situations, having a means to stay warm and sheltered is paramount:

- **Emergency Shelter:** A lightweight, compact shelter, such as a Mylar emergency tent or space blanket, offers protection from the elements.
- **Sleeping Bag:** Opt for a lightweight, insulated sleeping bag to retain body heat during cold nights.
- **Tarp or Poncho:** A versatile tarp or poncho can serve as an additional shelter, a ground cover, or a makeshift rain cover.

Water and Hydration

Securing a clean water source is critical for survival:

- **Water Filtration System:** A portable water filter or purification tablets allow you to convert potentially contaminated water into a safe, drinkable resource.
- **Collapsible Water Container:** Lightweight and space-saving, collapsible water containers facilitate efficient water storage and transportation.

Food and Nutrition

Pack nutrient-dense items that require minimal preparation:

- **Non-Perishable Foods:** Include energy bars, dehydrated meals, or vacuum-sealed snacks for a quick and efficient source of sustenance.
- **Compact Cooking Tools:** A lightweight, portable stove with fuel canisters or a compact camping cookware set enables you to prepare hot meals when necessary.

Navigation and Communication

Staying oriented and connected is crucial in survival situations:

- **Map and Compass:** A detailed map and a reliable compass are essential tools for navigation, especially in unfamiliar terrain.
- **Emergency Whistle:** A loud whistle can be a valuable signaling device for attracting attention or indicating your location.
- **Emergency Radio:** A battery-powered or hand-crank emergency radio allows you to stay informed about weather conditions and emergency broadcasts.

First Aid and Medical Supplies

Prioritize health and well-being by including a comprehensive first aid kit:

- **Bandages and Dressings:** Various sizes of bandages, sterile gauze, and adhesive dressings cater to different types of injuries.
- **Antiseptic Wipes:** Cleaning wounds is crucial, and antiseptic wipes help prevent infection.
- **Pain Relievers:** Include over-the-counter pain relievers for alleviating discomfort.
- **Prescription Medications:** If applicable, carry a supply of any necessary prescription medications.

Tools and Multi-Function Items

Versatile tools can be invaluable in a variety of situations:

- **Multi-Tool:** A multi-tool combines essential functions like cutting, gripping, and screwdriving in one compact device.
- **Knife:** A reliable, sharp knife serves various purposes, from food preparation to emergency situations.
- **Duct Tape:** Duct tape is a versatile fix-all that can be used for repairs, makeshift bandages, or securing items.

Illumination

Maintain visibility in low-light conditions with dependable lighting tools:

- **Headlamp or Flashlight:** Choose a durable, long-lasting headlamp or flashlight with extra batteries.
- **Chemical Light Sticks:** Chemical light sticks provide a reliable, portable light source without requiring batteries.

Clothing and Protection

Include weather-appropriate clothing and protective gear:

- **Weather-Resistant Clothing:** Pack extra layers, waterproof jackets, and durable footwear suitable for the climate.
- **Hat and Gloves:** Protect yourself from the elements with a hat for shade or warmth and gloves for handling various tasks.

Personal Hygiene Items

Maintaining personal hygiene is essential for overall well-being:

- **Travel-Sized Toiletries:** Include travel-sized toothpaste, soap, and personal hygiene items to stay clean.
- **Hand Sanitizer:** Keep a small bottle of hand sanitizer to minimize the risk of infections.

Personal Documents and Identification

Store essential documents in a secure, waterproof container:

- **Identification:** Include copies of identification, emergency contact information, and relevant documents.
- **Cash:** Carry a small amount of cash in case electronic payment methods are unavailable.

Chapter 3:

First Aid Basics

First aid basics are a set of essential skills and protocols that anyone can learn to provide immediate care for common injuries. Whether at home, in the workplace, or in outdoor settings, having a foundational understanding of first aid is crucial for promoting the well-being of individuals in need. Let's explore the key principles and treatment protocols for addressing common injuries.

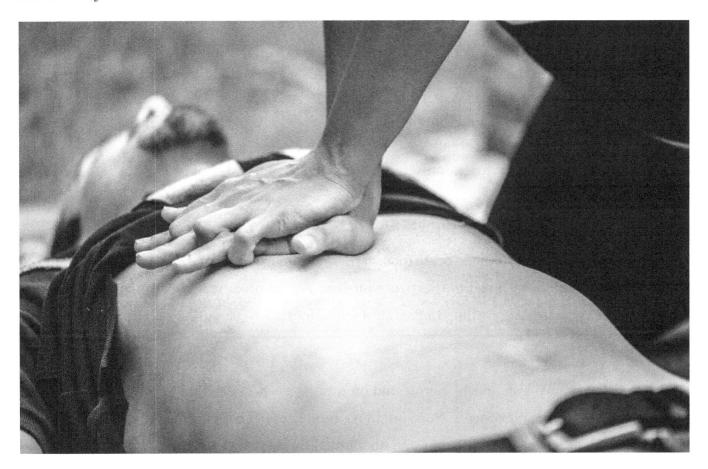

1. **Assessing the Scene**

The first step in any first aid situation is to ensure the safety of both the rescuer and the injured individual. Assess the scene for potential hazards, and take measures to eliminate or minimize risks before approaching the injured person.

2. Checking for Responsiveness

Determine if the individual is responsive by tapping them gently and asking if they are okay. If there is no response, shout for help, or call emergency services if available.

3. Activating Emergency Services

In situations involving serious injuries, activate emergency services immediately by calling the local emergency number. Provide clear and concise information about the nature of the emergency and the location.

4. ABCs of First Aid

The ABCs—Airway, Breathing, and Circulation—are critical components of assessing and addressing emergencies.

- **Airway:** Ensure the airway is clear by tilting the head back slightly and lifting the chin. If an obstruction is visible, remove it.
- **Breathing:** Check for breathing by observing the chest rise and fall. If the person is not breathing, initiate rescue breaths.
- **Circulation:** Check for a pulse. If there is no pulse, begin CPR (cardiopulmonary resuscitation).

5. CPR (Cardiopulmonary Resuscitation)

When a person's breathing or heartbeat stops, cardiopulmonary resuscitation (CPR) is a technique that can save their life. Involved in this procedure are rescue breaths and chest compressions. If trained in CPR, perform it according to the latest guidelines.

6. Choking

The Heimlich maneuver should be performed on a person who is choking and has no ability to breathe or speak during the process. You should position yourself behind the individual, wrap your arms around their waist, and perform rapid abdominal thrusts until the thing that is creating the obstruction is brought out of their body.

7. Bleeding and Wound Care

For bleeding injuries, take the following steps:

- **Apply Pressure:** Use a clean cloth or bandage to apply direct pressure on the wound.

- **Elevate the Injured Limb:** If possible, elevate the injured area to reduce blood flow.
- **Use a Tourniquet:** If bleeding is severe and uncontrollable, consider using a tourniquet above the wound, but ensure it is a last resort.

8. **Burns**

Treat minor burns by:

- **Cooling the Burn:** Gently run lukewarm water over the burn for 10-20 minutes to alleviate pain and minimize swelling.
- **Covering with a Clean Cloth:** Apply a sterile, non-adhesive bandage or a clean cloth to cover the burn.
- **Pain Management:** Administer over-the-counter pain relievers if necessary.

9. **Fractures and Sprains**

When dealing with fractures or sprains:

- **Immobilize the Injured Area:** Prevent further movement by immobilizing the injured limb using a splint or bandage.
- **Apply Ice:** Use ice packs to reduce swelling, applying them for 15-20 minutes at a time.
- **Elevate the Injured Limb:** If possible, elevate the injured area to minimize swelling.

10. **Head Injuries**

Head injuries require careful attention:

- **Stabilize the Head and Neck:** Never move the person's head or neck if you suspect that they have suffered an injury to their head or neck. Ensure that they remain stable in their current position.
- **Apply Cold Compress:** Use a cold compress to reduce swelling, applying it for short periods.

11. **Allergic Reactions**

For mild allergic reactions:

- **Administer Antihistamines:** Give antihistamines to reduce mild allergy symptoms.
- **Use an EpiPen:** If someone has a known severe allergy and is experiencing anaphylaxis, use an epinephrine auto-injector (EpiPen) if available.

12. **Seizures**

During a seizure:

- **Clear the Area:** Remove any nearby objects that could cause harm during the seizure.
- **Protect the Head:** Place the person on their side to protect their head and maintain an open airway.
- **Do Not Restrain:** Do not restrain the person during the seizure, and stay with them until it subsides.

13. **Shock**

Shock can occur after a severe injury or trauma:

- **Lay the Person Down:** Have the person lie down and elevate their legs to improve blood flow.
- **Cover with a Blanket:** Keep the person warm by covering them with a blanket.
- **Do Not Offer Food or Drink:** Avoid giving food or drink to someone in shock.

14. **Eye Injuries**

For eye injuries:

- **Flush with Water:** Rinse the eye with cool, clean water for at least 15 minutes to remove irritants.
- **Do Not Rub the Eye:** Avoid rubbing the injured eye, as it may cause further damage.
- **Cover with a Loose Bandage:** If there is an object embedded, cover the eye with a loose bandage and seek medical help.

15. **Dehydration**

To address dehydration:

- **Oral Rehydration:** Encourage the person to drink oral rehydration solutions, water, or clear fluids.
- **Avoid Caffeine and Alcohol:** Caffeine and alcohol can contribute to dehydration, so it's best to avoid them.

Chapter 4:

Essential Medicine Inventory

Building an essential medicine inventory is a prudent and responsible measure to ensure access to necessary medications in various situations. Whether creating a home medicine cabinet, preparing for travel, or establishing a comprehensive first aid kit, having a curated selection of key medicines can address common health concerns and emergencies.

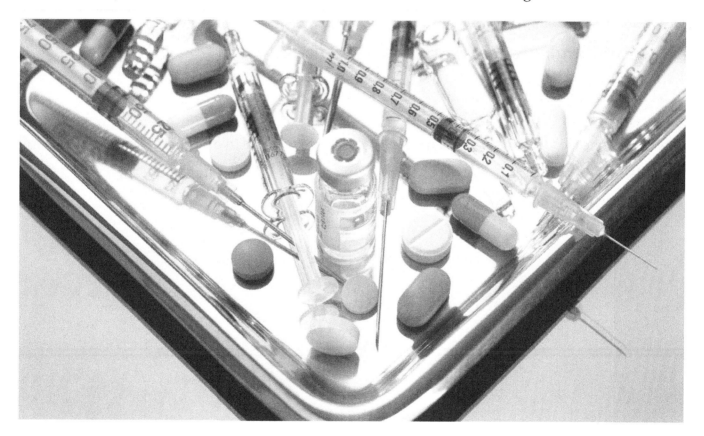

Pain Relief and Fever Reducers

- **Acetaminophen (Tylenol):** Effective for reducing pain and fever, acetaminophen is a common over-the-counter medication. It is suitable for various conditions, including headaches, muscle aches, and mild to moderate pain.
- **Ibuprofen (Advil, Motrin):** Ibuprofen is a particular type of nonsteroidal anti-inflammatory medicine (NSAID) that is effective in alleviating pain, lowering fever, and reducing inflammation. It is particularly effective for conditions such as muscle sprains, menstrual cramps, and headaches.

Antihistamines

- **Diphenhydramine (Benadryl):** An antihistamine that alleviates symptoms of allergies, including itching, sneezing, and hives. It can also be used as a sleep aid in some cases.
- **Cetirizine (Zyrtec) or Loratadine (Claritin):** These antihistamines are non-drowsy alternatives for allergy relief and are suitable for long-term use.

Gastrointestinal Medications

- **Loperamide (Imodium):** A medication for the treatment of diarrhea. It helps reduce the frequency and urgency of bowel movements.
- **Antacids (Tums, Rolaids):** Antacids provide relief from heartburn and indigestion by neutralizing stomach acid.

Adhesive Bandages and Antiseptic Ointment

- **Adhesive Bandages (Band-Aids):** Essential for covering and protecting minor cuts and wounds.
- **Antiseptic Ointment (Neosporin):** Applied to wounds to prevent infection and promote healing.

Thermometer

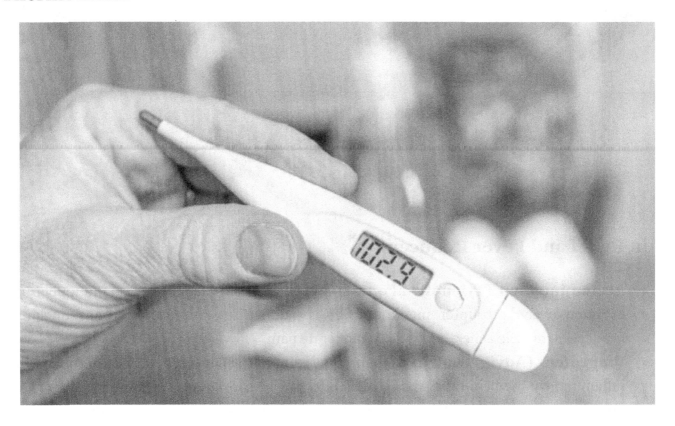

- **Digital Thermometer:** An essential tool for monitoring body temperature. A fever can be an indicator of various illnesses, and early detection is crucial.

Cough and Cold Medications

- **Cough Syrup or Lozenges:** Provide relief from cough symptoms.
- **Decongestants (Pseudoephedrine):** For nasal congestion, decongestants help reduce swelling in the nasal passages.
- **Expectorants (Guaifenesin):** These help thin mucus, making it easier to clear from the airways.

Oral Rehydration Solutions

- **Electrolyte Packets (e.g., Pedialyte):** Important for rehydration during episodes of dehydration caused by vomiting, diarrhea, or fever.

Allergy Medications

- **Epinephrine Auto-Injector (EpiPen):** For individuals with severe allergies, an epinephrine auto-injector can be life-saving in the event of an allergic reaction.
- **Prescription Allergy Medications:** For those with known allergies, having a supply of prescribed allergy medications is crucial.

Topical Corticosteroids

- **Hydrocortisone Cream:** Useful for treating skin irritations, itching, and mild rashes.

Oral Pain Relief

- **Orajel or Benzocaine Gel:** Provides relief from toothaches or oral pain.

Eye Drops

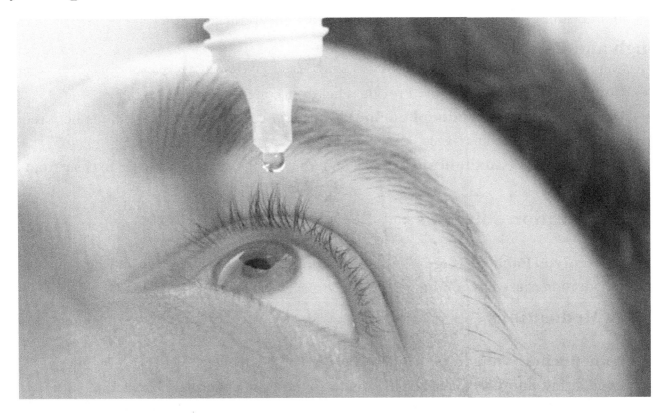

- **Artificial Tears:** Lubricating eye drops are helpful for dry or irritated eyes.

Motion Sickness Medications

- **Dimenhydrinate (Dramamine):** An over-the-counter medication for preventing and treating motion sickness.

Anti-fungal Cream

- **Clotrimazole or Miconazole:** Used for treating fungal infections, such as athlete's foot or yeast infections.

Prescription Medications

- **Prescription Medications:** Individuals with chronic conditions should keep an ample supply of their prescribed medications, including those for hypertension, diabetes, or other ongoing health issues.

Personalized Medications

- **Personalized Medications:** Consider any specialized medications required for individual health needs, such as asthma inhalers, epinephrine injectors for severe allergic reactions, or medications for chronic conditions.

Storage and Safety Tips:

1. **Check Expiration Dates:** Regularly review and replace medications that have expired.

2. **Keep Medications Secure:** Store medications in a cool, dry place away from direct sunlight. Keep them out of reach of children.

3. **Label Medications:** Clearly label medications, especially if they require specific instructions or if multiple family members share the same space.

4. **Create a Medication List:** Maintain a list of all medications in the inventory, including dosage instructions and potential side effects.

5. **Consult a Healthcare Professional:** If uncertain about the suitability of any medication or its potential interactions, consult with a healthcare professional.

Chapter 5:

Mental Health Management

Mental health management during emergencies is a crucial aspect of overall well-being. Stress, fear, and uncertainty can significantly impact mental health, and having effective coping strategies is essential for navigating challenging situations. Whether facing natural disasters, personal crises, or global emergencies, individuals can benefit from proactive mental health management techniques. Here are key approaches to coping with stress and fear during emergencies:

1. **Acknowledge and Validate Emotions:** The first step in mental health management is recognizing and acknowledging emotions. It's okay to feel fear, stress, or anxiety during emergencies. Validating these emotions is an important aspect of self-awareness.

2. **Stay Informed, but Limit Exposure:** While staying informed about the emergency is crucial, constant exposure to news and updates can contribute to heightened anxiety. Set specific times to check for updates and limit exposure to distressing information.

3. **Establish a Routine:** Creating a daily routine can provide a sense of stability and predictability during uncertain times. Include activities that bring comfort and relaxation, helping to maintain a sense of normalcy.

4. **Connect with Others:** Social support is invaluable for mental health. Reach out to friends, family, or support groups. Share your feelings, thoughts, and concerns. Connection with others provides a sense of belonging and shared experience.

5. **Engage in Mindfulness and Relaxation Practices:** Incorporating mindfulness and relaxation methods, such as engaging in deep breathing exercises, practicing meditation, or participating in yoga, can effectively reduce stress and foster a peaceful state of mind. These approaches are convenient and adaptable, making them suitable for any location.

6. **Physical Activity:** Regular participation in physical activities not only enhances physical well-being but also contributes positively to mental health. Exercise triggers the release of endorphins, leading to an improved mood and a reduction in stress levels.

7. **Set Realistic Goals:** Break down tasks into manageable goals. Setting realistic and achievable objectives can provide a sense of accomplishment, boosting confidence during challenging times.

8. **Maintain Healthy Habits:** Prioritize self-care by maintaining healthy habits. Make sure to get sufficient sleep, maintain a well-balanced diet, and steer clear of excessive intake of stimulants such as caffeine and alcohol. A well-nourished body contributes to a resilient mind.

9. **Seek Professional Help:** When stress and fear become too much to handle, reaching out for professional assistance reflects resilience. Mental health experts can offer personalized guidance, support, and coping strategies to address individual needs.

10. **Focus on What You Can Control:** During emergencies, it's common to feel a lack of control. Focus on aspects of your life that you can control, such as your reactions, daily routines, and self-care practices. This can empower you in the face of uncertainty.

11. **Engage in Hobbies and Activities:** Pursuing hobbies and activities you enjoy can be a positive distraction. Whether it's reading, crafting, or playing a musical instrument, engaging in activities that bring joy contributes to mental well-being.

12. **Practice Gratitude:** Cultivate a mindset of gratitude by acknowledging positive aspects of your life. Expressing gratitude can shift focus from challenges to the things that bring joy and appreciation.

13. **Build Resilience:** Develop the capacity to recover from challenges and setbacks. Enhance resilience by embracing change, gaining insights from past experiences, and fostering a positive mindset. The construction of resilience is a continuous journey that fortifies mental well-being as it progresses.

14. **Develop a Safety Plan:** In times of crisis, having a safety plan can provide a sense of security. This plan may include emergency contacts, coping strategies, and steps to take if stressors escalate.

15. **Foster a Sense of Purpose:** Identify and nurture a sense of purpose. This could involve contributing to community efforts, volunteering, or pursuing personal goals. Having a purpose can provide motivation and a sense of direction.

BOOK 6:
THE PERFECT PREPPER'S SURVIVAL CHECKLIST

Chapter 6:

Survival Checklist

Survival preparedness is a multifaceted endeavor that requires careful planning and consideration. This serves as a comprehensive guide, detailing essential items and actions to ensure thorough preparedness for a variety of scenarios. This checklist is designed to cover various aspects of survival, from basic necessities to specialized tools, with a focus on adaptability to different situations.

Water Supply

- **Portable Water Filtration System:** A compact and portable water filtration system allows you to convert potentially contaminated water into a safe, drinkable resource.
- **Water Storage Containers:** Include sturdy containers for storing an ample supply of water. Aim for at least one gallon per person per day.
- **Water Purification Tablets:** As a backup to filtration systems, water purification tablets can be effective in making water safe for consumption.

Food and Nutrition

- **Non-Perishable Foods:** Stock up on a variety of non-perishable foods, including canned goods, dried fruits, nuts, and energy bars.
- **Dehydrated Meals:** Lightweight and with a long shelf life, dehydrated meals are convenient for emergency situations.
- **Manual Can Opener:** Ensure you have a reliable manual can opener to access canned foods.
- **Compact Camping Stove:** A portable camping stove provides a means to cook food during emergencies. Include fuel canisters suitable for the stove.

Shelter and Warmth

- **Emergency Tent or Tarp:** Lightweight emergency tents or tarps can serve as temporary shelters, providing protection from the elements.
- **Sleeping Bag:** Invest in a durable and insulated sleeping bag to retain body heat in varying temperatures.

- **Mylar Space Blankets:** These compact and reflective blankets help retain body heat and are useful in emergency situations.

First Aid and Medical Supplies

- **Comprehensive First Aid Kit:** Bandages, antiseptic wipes, pain medications, and other critical medical supplies should be included in a first aid kit that is packed with all of the necessary items.
- **Prescription Medications:** Ensure an ample supply of any necessary prescription medications for all family members.
- **Medical Gloves and Masks:** Disposable gloves and masks provide protection against the spread of infections.
- **Basic Over-the-Counter Medications:** Include medications for common ailments such as pain relievers, antacids, and anti-diarrheal medication.

Tools and Multi-Function Items

- **Multi-Tool:** A versatile multi-tool combines various functions, including cutting, gripping, and screwdriving.
- **Knife:** A reliable and sharp knife is indispensable for various tasks, including food preparation and emergency situations.
- **Flashlight with Extra Batteries:** Ensure you have a durable flashlight with extra batteries for illumination during power outages.
- **Duct Tape:** This versatile tool can be used for repairs, makeshift bandages, or securing items.

Communication Devices

- **Battery-Powered or Hand-Crank Emergency Radio:** Stay informed about weather conditions and emergency broadcasts.
- **Whistle:** A loud whistle serves as a signaling device and can be crucial for attracting attention.
- **Portable Power Bank:** Charge electronic devices with a portable power bank to stay connected in emergency situations.

Personal Hygiene Items

- **Travel-Sized Toiletries:** Include travel-sized toothpaste, soap, and other personal hygiene items to maintain cleanliness.
- **Hand Sanitizer:** Keep a small bottle of hand sanitizer to minimize the risk of infections.

- **Feminine Hygiene Products:** For female preppers, ensure an adequate supply of feminine hygiene products.

Clothing and Protection

- **Weather-Appropriate Clothing:** Pack extra layers, waterproof jackets, durable footwear, and weather-appropriate clothing.
- **Hat and Gloves:** Protect yourself from the elements with a hat for shade or warmth and gloves for various tasks.
- **N95 Masks:** In situations where air quality may be compromised, N95 masks offer respiratory protection.

Financial and Legal Documents

- **Copies of Important Documents:** A container that is watertight should be used to store duplicates of important documents such as your identification, insurance policies, medical records, and other important paperwork.
- **Cash:** Maintain a small amount of cash in various denominations, as ATMs and electronic payment methods may be unavailable.

Self-Defense and Security

- **Personal Protection Items:** Depending on legal regulations, consider items such as pepper spray, personal alarms, or self-defense tools.
- **Firearms and Ammunition:** If legally permitted and trained, having a firearm with sufficient ammunition can enhance personal security.

Navigation and Maps

- **Map and Compass:** A detailed map and a reliable compass are crucial for navigation, especially in unfamiliar terrain.
- **GPS Device:** A battery-powered GPS device can provide additional navigation assistance.

Entertainment and Comfort

- **Books, Games, or Entertainment Devices:** Include items that provide entertainment and alleviate stress during extended periods of confinement.
- **Comfort Items:** Pack personal comfort items, such as a favorite blanket or a small pillow.

Extra Supplies for Children and Pets

- **Baby Supplies:** For families with infants, include extra diapers, formula, and baby essentials.
- **Pet Supplies:** If you have pets, ensure you have extra food, water, and any necessary medications for them.

Maintenance and Repair Items

- **Basic Tools:** Include a set of basic tools for minor repairs and maintenance.
- **Rope and Cordage:** Durable ropes and cordage are versatile for securing items or constructing makeshift shelters.

Mental Health Support

- **Stress-Relief Items:** Include items such as a journal, stress balls, or comforting items to support mental well-being.
- **Contact Information for Mental Health Professionals:** Have contact information for mental health professionals in case additional support is needed.

BOOK 7:
WATER MASTERY

Chapter 1:

Water Sourcing Techniques

Water is a fundamental resource for survival, and knowing innovative water sourcing techniques can be essential in various situations. From emergency scenarios to outdoor adventures, the ability to find water is a skill that can make a significant difference.

1. Natural Indicators

Observing the surrounding environment for natural indicators can be a valuable technique. Look for signs such as:

- **Vegetation:** Lush and green vegetation, especially in arid areas, may indicate the presence of water sources underground.
- **Animal Activity:** Wildlife, birds, and insects tend to congregate around water sources. Observing their behavior can lead you to nearby water.
- **Topography:** Understanding the terrain and identifying depressions, valleys, or low-lying areas can suggest potential water catchments.

2. Solar Still Construction

A solar still is a simple yet effective way to extract water from the soil, even in arid conditions. To construct a solar still:

- **Dig a Hole:** Dig a hole in the ground, ensuring it's deep enough to collect water.
- **Place a Container:** Place a container at the center of the hole to collect condensation.
- **Cover with Plastic:** Cover the hole with clear plastic, securing the edges with rocks or soil.
- **Weight the Plastic:** Place a small rock or weight at the center of the plastic above the container.

As the sun heats the ground, moisture evaporates, condenses on the plastic, and drips into the container.

3. **Water from Plants**

Certain plants, particularly in humid environments, can serve as a source of water. Techniques include:

- **Transpiration Bags:** Tie a plastic bag around a leafy branch, creating a transpiration bag. Over time, the plant releases moisture that collects in the bag.
- **Plant Stems:** Some plant stems, like those of certain cacti, can be cut open to access water stored inside.

4. **Rock Moisture**

In areas with dew or high humidity, rocks can be a source of moisture. To extract water from rocks:

- **Choose Porous Rocks:** Select rocks that are porous and likely to absorb moisture.
- **Expose to Sun:** Place the rocks in direct sunlight to heat them.
- **Collect Condensation:** As the rocks cool in the evening, they may release condensation, which can be collected.

5. **Bamboo Water Collection**

Bamboo is known for its water-collecting properties. To use bamboo for water sourcing:

- **Cut Bamboo at an Angle:** Cut a mature bamboo stalk at a slight angle, leaving one end lower.
- **Position Upright:** Place the bamboo in an upright position, allowing water to collect at the lower end.
- **Harvest Water:** Periodically check and collect water that accumulates in the lower end.

6. **Underground Water Detection**

In some situations, water may be found underground. Techniques for detecting underground water include:

- **Divining Rods:** Some people use divining rods, often Y-shaped branches, to detect underground water. The rods supposedly react when held over a water source.
- **Listen for Sounds:** Listen for the sound of running water or groundwater flow, especially in quiet surroundings.
- **Observing Animal Behavior:** Animals digging or congregating in specific areas may indicate access to underground water.

7. Fog Nets

In foggy or coastal areas, harnessing moisture from the air can be achieved with fog nets. These nets capture tiny water droplets from the fog, providing a water source. This method is often used in regions with low rainfall.

8. Inflatable Solar Stills

Similar to traditional solar stills, inflatable solar stills are portable and lightweight. They use sunlight to create a greenhouse effect, trapping and collecting moisture from the soil inside the device.

9. Atmospheric Water Generators

Advanced technology has led to the development of atmospheric water generators. These devices extract moisture from the air, condensing it into drinkable water. While often used in modern settings, smaller and portable versions are available for personal use.

10. Ice and Snow Melting

In cold climates, ice and snow can be melted to obtain water. Techniques include:

- **Solar Melting:** Use sunlight to melt ice or snow by placing it in a container or on a reflective surface.
- **Body Heat:** Place ice or snow in a container and use body heat to facilitate melting.
- **Heated Rocks:** Warm rocks in a fire and use them to melt snow or ice in a container.

Chapter 2:

DIY Purification Systems

Creating your own do-it-yourself (DIY) water purification system is a practical and empowering skill, especially in situations where clean water is scarce or unavailable. This step-by-step purification guide will take you through the process of constructing a simple yet effective DIY water purification system using readily available materials. Whether you're an outdoor enthusiast, a prepper, or facing an emergency situation, having the knowledge to purify water can be invaluable for ensuring a safe and reliable water supply.

Materials Needed:

- **Large Container or Bucket:** Start with a clean and preferably large container or bucket. This will be the primary vessel for holding and treating the water.
- **Clean Cloth or Coffee Filter:** Use a clean cloth or coffee filter to remove larger particles and debris from the water. This initial filtration step helps prevent clogging of finer filters and enhances the effectiveness of subsequent purification methods.
- **Sand:** Fine sand acts as a natural filter, removing impurities and sediments from the water. Ensure the sand is clean and free from contaminants.
- **Activated Charcoal:** Activated charcoal is highly effective in adsorbing impurities, chemicals, and odors from water. You can obtain activated charcoal from stores or create it by heating charcoal in the presence of a gas that causes it to become porous.
- **Gravel or Small Rocks:** Gravel or small rocks provide additional filtration, helping to remove larger particles and improve the overall clarity of the water.
- **UV Light Source (Optional):** If available, a UV light source can be used to further disinfect the water by neutralizing bacteria, viruses, and other harmful microorganisms. This step is optional but adds an extra layer of protection.

Step-by-Step DIY Water Purification:

Step 1: Prepare the Container: Ensure the container is clean and free from any contaminants. It should be large enough to hold the entire purification system and provide ample space for water treatment.

Step 2: Initial Filtration: Place a clean cloth or coffee filter over the opening of the container. This will help remove larger particles, leaves, and debris from the water. Secure the cloth in place, forming a makeshift filter.

Step 3: Layer of Gravel or Small Rocks: Add a layer of gravel or small rocks to the container. This layer acts as the first stage of filtration, capturing larger particles and improving the water's overall clarity.

Step 4: Layer of Sand: Pour a layer of fine sand over the gravel. The sand will further filter the water, removing smaller particles and impurities.

Step 5: Layer of Activated Charcoal: Add a layer of activated charcoal on top of the sand. The activated charcoal will adsorb chemicals, contaminants, and odors, enhancing the water's purity.

Step 6: Repeat Layers if Necessary: Depending on the container's size and the level of water purification required, you can repeat the layers of gravel, sand, and activated charcoal. This step is especially useful if you have a larger container or need to purify a significant amount of water.

Step 7: Allow Settling Time: Allow the water to pass through the layers of filtration media and settle in the container. This settling time allows the purification process to take effect.

Step 8: Additional UV Treatment (Optional): If you have a UV light source, you can use it to further disinfect the water. Shine the UV light into the container, ensuring that it reaches all parts of the water. This step helps neutralize harmful microorganisms.

Step 9: Collect Purified Water: Once the water has undergone the purification process, carefully collect it from the container. You can use a clean container or a water dispenser for this purpose.

Step 10: Boiling (Optional): For an extra layer of precaution, especially in emergency situations, consider boiling the purified water before consumption. Boiling kills most microorganisms and ensures the water is safe for drinking.

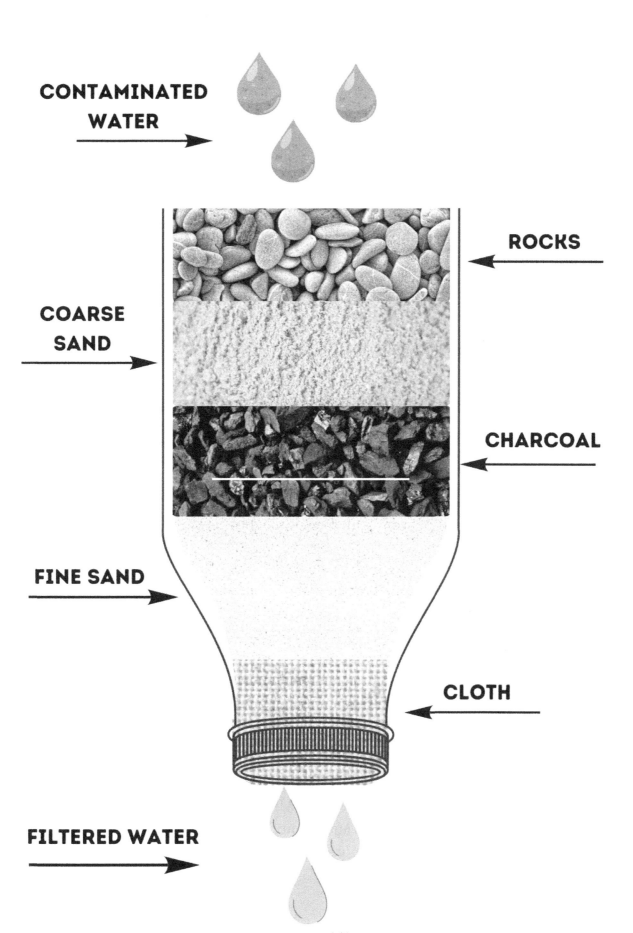

CONTAMINATED WATER

ROCKS

COARSE SAND

CHARCOAL

FINE SAND

CLOTH

FILTERED WATER

Chapter 3:

Water Storage Solutions

Water storage is a critical aspect of preparedness, ensuring a stable and reliable water supply in various scenarios, from emergencies to off-grid living. Implementing effective water storage solutions involves both long-term planning and conservation strategies.

Long-Term Water Storage Solutions

1. **Clean and Durable Containers:** Choose containers made of food-grade materials that are specifically designed for long-term water storage. Options include high-density polyethylene (HDPE) containers, water barrels, or water storage tanks. Ensure the containers are clean and free from contaminants before use.

2. **Dark-Colored Containers:** Opt for containers that are dark or opaque in color. This helps prevent sunlight from penetrating the containers, reducing the growth of algae and minimizing the risk of bacterial contamination. Sunlight exposure can also break down the quality of stored water over time.

3. **Sealed and Airtight Lids:** Ensure that the storage containers have sealed and airtight lids. This prevents external contaminants from entering and maintains the water's quality. Sealed lids also prevent evaporation, preserving the stored water for an extended period.

4. **Water Preservatives:** To enhance the longevity of stored water, consider adding water preservatives or stabilizers. These additives help prevent the growth of algae, bacteria, and other microorganisms. Follow the manufacturer's guidelines for the appropriate amount of preservative based on the volume of water.

5. **Rotation System:** Implement a rotation system for your water storage. Regularly use and replace stored water to ensure freshness and quality. This practice also allows you to inspect containers for any signs of damage or degradation.

6. **Underground Cisterns:** For a more permanent solution, consider installing underground cisterns. These large tanks are buried underground, protecting stored water from temperature fluctuations and sunlight. Underground storage helps maintain a consistent temperature, reducing the risk of bacterial growth.

Water Conservation Strategies

1. **Rainwater Harvesting:** Harvesting rainwater is an eco-friendly method that can significantly contribute to water conservation. Set up rain barrels or larger cisterns to collect rainwater from roofs. Install gutters to direct rainwater into these storage containers.

2. **Greywater Systems:** Implement greywater systems to recycle and reuse water from activities such as laundry, dishwashing, and bathing. Greywater, when properly treated, can be used for irrigation, flushing toilets, or other non-potable purposes, reducing the demand on fresh water sources.

3. **Water-Efficient Appliances:** Upgrade to water-efficient appliances, including low-flow toilets, faucets, and showerheads. These appliances help reduce water consumption without compromising functionality.

4. **Drip Irrigation:** Employ drip irrigation systems for your gardens and plants. These systems directly supply water to the roots, reducing water wastage and fostering effective water utilization. Mulching around plants also helps retain soil moisture.

5. **Fix Leaks Promptly:** Repair any leaks in plumbing or irrigation systems promptly. Even small leaks can lead to significant water wastage over time. Regularly check for and address leaks to conserve water and prevent unnecessary expenses.

6. **Watering Schedule:** The individual requirements of each variety of plant should be taken into consideration when developing a watering regimen for outdoor plants of any kind. Watering during the early morning or late evening reduces evaporation, ensuring more effective water absorption by the soil.

7. **Educational Initiatives:** Promote water conservation within your community through educational initiatives. Raise awareness about the importance of responsible water use and share tips on efficient water management.

8. **Mulching and Ground Cover:** When you apply mulch to garden beds, you can help the soil retain moisture and reduce the amount of times you need to water them. Ground cover plants also help prevent evaporation and maintain soil hydration.

9. **Reclaimed Water for Landscaping:** In areas where reclaimed water is available, consider using it for landscaping and irrigation. Reclaimed water is treated wastewater that can be repurposed for non-potable uses, conserving fresh water for essential needs.

10. **Native and Drought-Tolerant Plants:** Choose native plants and those adapted to the local climate, as they often require less water. Drought-tolerant landscaping reduces water demand and promotes sustainable water use.

BOOK 8: SUSTENANCE AND NUTRITION

Chapter 1:

Foraging Guide

Venturing into the world of foraging opens up a treasure trove of sustenance provided by nature. This foraging guide aims to equip you with the knowledge to identify edible plants and insects, unlocking the potential of diverse and nourishing resources available in the wild. Whether you're an outdoor enthusiast, a survivalist, or someone curious about sustainable living, understanding the edible offerings of nature is a valuable skill.

Edible Plants

Dandelion (Taraxacum officinale)

- **Identification:** Recognizable by its distinctive yellow flowers and toothed leaves. All parts of the dandelion are edible, including leaves, flowers, and roots.
- **Uses:** Leaves can be eaten raw or cooked, flowers can be used in salads or as a garnish, and roots can be roasted for a coffee substitute.

Wild Garlic (Allium ursinum)
- **Identification:** Characterized by its long, pointed leaves and clusters of star-shaped white flowers. Crushed leaves emit a strong garlic aroma.
- **Uses:** Leaves can be used in salads, pesto, or cooked dishes. Bulbs can be harvested for a mild garlic flavor.

Stinging Nettle (Urtica dioica)
- **Identification:** Has serrated leaves and tiny, hair-like structures that can cause skin irritation.
- **Uses:** Young leaves can be cooked and eaten, providing a good source of vitamins and minerals. Cooking neutralizes the stinging hairs.

Chickweed (Stellaria media)
- **Identification:** Small, star-like white flowers and paired leaves.
- **Uses:** Young shoots and leaves are edible and can be added to salads or cooked as a nutritious green.

Plantain (Plantago major)
- **Identification:** Broad, ribbed leaves in a rosette formation. Commonly found in lawns and disturbed soil.
- **Uses:** Leaves can be eaten raw or cooked, and they have a mild, nutty flavor. Known for potential medicinal properties.

Wood Sorrel (Oxalis spp.)
- **Identification:** Clover-like leaves with a sour taste.
- **Uses:** Leaves and flowers are edible and add a citrusy flavor to salads or can be eaten on their own.

Burdock (Arctium lappa)
- **Identification:** Large leaves and burr-like seed heads.
- **Uses:** Young leaves can be cooked, and the root can be harvested, peeled, and eaten. Known for its potential health benefits.

Wild Strawberries (Fragaria vesca)
- **Identification:** Similar to cultivated strawberries but smaller with a more intense flavor.
- **Uses:** Berries are sweet and can be eaten raw or used in various culinary applications.

Cattail (Typha spp.)
- **Identification:** Long, blade-like leaves and cylindrical flower spikes.
- **Uses:** Young shoots, rhizomes, and pollen can be eaten. Rhizomes can be cooked, and the pollen can be used as a flour supplement.

Mallow (Malva spp.)
- **Identification:** Heart-shaped leaves and small, pink or white flowers.

- **Uses:** Leaves and flowers are edible and can be used in salads or cooked dishes.

Acorns (Quercus spp.)
- **Identification:** Nuts from oak trees, usually with a cap covering part of the nut.
- **Uses:** Acorns can be processed to remove bitter tannins and then ground into flour for various culinary applications.

Pine Needle Tea (Pinus spp.)
- **Identification:** Long needles arranged in clusters on pine branches.
- **Uses:** Needles can be used to make a vitamin C-rich tea. Avoid poisonous yew and Norfolk Island pine; opt for edible pine species.

Edible Insects

Crickets

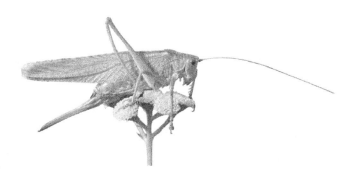

- **Identification:** Six legs, two antennae, and wings. Brown or black coloration.
- **Uses:** Crickets can be roasted or ground into a powder for use in recipes. Rich in protein.

Mealworms (Darkling Beetle Larvae)
- **Identification:** Small, segmented larvae with a golden-brown color.
- **Uses:** Can be eaten raw, roasted, or ground into flour. High in protein and fats.

Grasshoppers
- **Identification:** Long hind legs for jumping, large eyes, and wings.
- **Uses:** Can be roasted or cooked and eaten. Rich in protein and commonly found in various ecosystems.

Ants
- **Identification:** Six legs, segmented body, and elbowed antennae.
- **Uses:** Can be eaten raw or added to dishes. Some ants have a citrusy flavor.

Termites
- **Identification:** Small, pale insects with soft bodies.
- **Uses:** Can be consumed raw or cooked. A good source of protein.

Silkworm Pupae
- **Identification:** Creamy-white or brown pupae with a soft exterior.
- **Uses:** Often eaten roasted or boiled. Considered a delicacy in some cultures.

Beetle Larvae (e.g., Rhinoceros Beetle Grubs)
- **Identification:** Grub-like larvae with a soft body.
- **Uses:** Can be cooked and eaten. High in protein and commonly consumed in certain regions.

Butterfly and Moth Larvae (Caterpillars)
- **Identification:** Soft, segmented larvae with distinct body markings.
- **Uses:** Can be cooked and eaten. Some caterpillars are consumed in various cuisines.

Dragonfly and Damselfly Nymphs
- **Identification:** Aquatic larvae with a distinctive, elongated body.
- **Uses:** Can be cooked and eaten. Commonly found in freshwater habitats.

Honeybee Larvae
- **Identification:** Soft, white larvae found in honeycomb.
- **Uses:** Considered a delicacy in some cultures. Can be eaten raw or cooked.

Safety Considerations

- Always positively identify plants and insects before consumption.
- Avoid areas contaminated with pesticides or pollutants.
- Be cautious of potential allergic reactions, especially if trying a new food for the first time.
- Consult local foraging guides or experts for region-specific information.

Chapter 2:

Ethical Hunting Practices

Engaging in hunting is an ancient practice deeply intertwined with human history, providing sustenance, resources, and even cultural significance. As we explore ethical hunting practices, the emphasis is on fostering sustainability, respect for nature, and responsible stewardship. Ethical hunting involves a set of principles and techniques that prioritize the well-being of ecosystems, wildlife populations, and the ethical treatment of animals. This guide aims to illuminate the key aspects of ethical hunting, ensuring a balanced and respectful approach to this age-old pursuit.

Understanding Ethical Hunting

1. **Respect for Wildlife:** Ethical hunting begins with a profound respect for wildlife. Hunters should appreciate the intrinsic value of each species and recognize their role in maintaining a balanced ecosystem. This respect extends to understanding the behavior, habitats, and life cycles of the animals being pursued.

2. **Knowledge and Education:** A foundational element of ethical hunting is continuous learning. Hunters should invest time in acquiring knowledge about the species they hunt, their habitats, and the ecosystems they are part of. Staying informed about conservation practices and wildlife management contributes to responsible hunting.

3. **Compliance with Laws and Regulations:** Ethical hunters adhere strictly to hunting laws and regulations. These rules are designed to protect wildlife populations, ensure sustainable practices, and maintain the delicate balance of ecosystems. Knowledge of and compliance with these laws are essential for ethical hunting.

4. **Selective Harvesting:** Ethical hunters practice selective harvesting, targeting specific individuals within a population based on age, gender, or health. This approach helps maintain healthy population structures and genetic diversity while ensuring the sustainability of the overall population.

5. **Fair Chase:** The concept of fair chase emphasizes the pursuit of game in a manner that provides them with a reasonable chance of escape. Ethical hunters avoid practices that could

give them an unfair advantage, such as excessive use of technology or pursuing animals in confined spaces.

Sustainable Practices

1. **Population Management:** Ethical hunters actively contribute to population management efforts. They support wildlife agencies' efforts to regulate populations within ecological carrying capacities, preventing overpopulation and the subsequent strain on ecosystems.

2. **Habitat Conservation:** Hunters play a crucial role in habitat conservation. By supporting initiatives that protect and restore natural habitats, ethical hunters contribute to the well-being of entire ecosystems. Conserving these habitats ensures the availability of suitable environments for wildlife.

3. **Understanding Ecology:** A fundamental tenet of ethical hunting is an understanding of ecological processes. Hunters should recognize the interconnectedness of species, the impact of environmental changes, and the cascading effects that can result from alterations to ecosystems.

4. **Leave No Trace:** Ethical hunters adhere to the principle of "leave no trace." This means minimizing their impact on the environment by properly disposing of waste, avoiding unnecessary habitat disturbance, and respecting the wilderness they traverse.

5. **Supporting Conservation Initiatives:** Ethical hunters actively support conservation organizations and initiatives. They recognize the importance of financial contributions and volunteer efforts in preserving wildlife habitats, conducting research, and implementing conservation programs.

Respectful and Humane Practices

1. **Quick and Humane Kills:** Ethical hunters prioritize quick and humane kills. This involves using appropriate weaponry, maintaining proficiency in marksmanship, and understanding the anatomy of the animals they hunt to ensure a swift and ethical harvest.

2. **Minimizing Stress:** Ethical hunters aim to minimize stress on animals during the hunting process. This includes reducing unnecessary pursuit, ensuring efficient kills, and handling harvested game with care to minimize additional stress.

3. **Full Utilization of Harvested Animals:** Ethical hunters commit to utilizing the entire animal, minimizing waste. This includes harvesting meat, using hides or fur, and employing other parts of the animal for various purposes. This respect for the animal's life extends to making the most of its contributions.

4. **Ethical Trophy Hunting:** In cases of trophy hunting, ethical hunters emphasize the utilization of all parts of the animal, ensuring that the hunt serves conservation purposes and benefits local communities. This approach contrasts with trophy hunting solely for the sake of acquiring trophies.

5. **Integrity in Reporting:** Ethical hunters maintain integrity in reporting their hunting experiences. This includes accurate documentation of the hunt, adherence to bag limits, and truthful reporting of harvest data. Honest reporting contributes to effective wildlife management.

Personal Ethics and Responsibility

1. **Personal Reflection and Accountability:** Ethical hunters engage in personal reflection on their motivations, practices, and impact. They hold themselves accountable for their actions and continuously strive to improve their skills, knowledge, and ethical decision-making.

2. **Cultural and Community Considerations:** Ethical hunting recognizes the cultural and community aspects of the practice. It respects traditions and values while adapting to contemporary ethical standards. Ethical hunters actively engage with their communities to foster responsible hunting practices.

3. **Advocacy for Ethical Hunting:** Ethical hunters become advocates for responsible hunting within their communities. They share knowledge, promote ethical standards, and contribute to a positive and sustainable hunting culture.

Chapter 3:

Food Preservation Methods

Starting the adventure of food preservation unveils various ways to prolong the lifespan of perishable items and guarantee a consistent food stock. Advanced food preservation methods go beyond basic techniques, providing innovative approaches to store food for the long term.

Freeze Drying

Freeze drying is an advanced method that removes moisture from food, preventing the growth of bacteria, yeasts, and molds. In this process, food is frozen and then subjected to low pressure, causing the frozen water content to sublimate directly into vapor. The result is lightweight, shelf-stable food with retained nutritional value and flavor.

Applications: Freeze-dried fruits, vegetables, meats, and even complete meals are popular for their lightweight nature, extended shelf life, and ease of rehydration.

Fermentation

Fermentation is a time-honored method that involves the conversion of sugars and starches in food into alcohol or organic acids by microorganisms like bacteria, yeast, or molds. Not only does this method preserve the food, but it also improves its flavor and the nutritional composition of the meal. Common fermented foods include sauerkraut, kimchi, yogurt, and pickles.

Applications: Fermented foods contribute to gut health, and the process can be applied to various vegetables, dairy products, and even beverages like kombucha.

Canning with Pressure

Pressure canning is a method that uses high pressure to achieve elevated temperatures, effectively eliminating harmful microorganisms and spores that regular boiling water canning might not address. This technique is crucial for preserving low-acid foods like meats, poultry, and vegetables.

Applications: Pressure canning is widely used for preserving soups, stews, and a variety of low-acid foods in jars, ensuring safety and longevity.

Controlled Atmosphere Storage

This method involves adjusting and maintaining the atmospheric conditions in storage spaces to slow down the ripening and decay of fruits and vegetables. Oxygen, carbon dioxide, and humidity levels are carefully controlled to create an environment that prolongs the freshness of perishable items.

Applications: Controlled atmosphere storage is commonly employed in commercial settings for apples, pears, and other fruits to extend their storage life.

Dehydration with Technology

While traditional sun drying is a common dehydration method, advanced technologies like electric dehydrators offer more precise control over the drying process. These devices use consistent heat and airflow to remove moisture from foods, resulting in lightweight, shelf-stable products.

Applications: Dehydrators are ideal for preserving fruits, vegetables, herbs, and even making jerky from meats.

High-Pressure Processing (HPP)

The process of high-pressure processing includes applying high levels of hydrostatic pressure to food that has been properly packaged. This method inactivates spoilage microorganisms and pathogens, extending the shelf life of the product while preserving its nutritional quality and sensory attributes.

Applications: HPP is commonly used for juices, deli meats, and ready-to-eat meals, offering a method that maintains product quality without the need for high heat.

Vacuum Packing

Vacuum packing removes the air from the packaging, creating an environment where spoilage microorganisms cannot thrive. This method helps prevent oxidation and slows down the degradation of food, contributing to extended shelf life.

Applications: Vacuum packing is used for a variety of foods, including meats, cheeses, and dry goods, providing an airtight seal that maintains freshness.

Radiation Preservation

Ionizing radiation, such as gamma radiation or electron beam irradiation, is employed to kill bacteria, parasites, and pathogens in food. This process disrupts the DNA of microorganisms, preventing them from reproducing and causing spoilage.

Applications: Radiation preservation is used for spices, herbs, and certain dry food items, ensuring microbial safety without compromising the quality of the food.

Osmotic Dehydration

Osmotic dehydration involves immersing food in a hypertonic solution, drawing out moisture from the food. This method not only preserves the food but also imparts sweetness and enhances flavor. Common solutions include sugar or salt solutions.

Applications: Osmotic dehydration is applied to fruits, particularly in the production of candied fruits, preserving their texture and taste.

SEASON	MONTHS	FRUITS	VEGETABLES
Spring	March - May	Pineapple - Strawberries - Mango Rhubarb - Lychee - Guava - Kumquat	Asparagus - Carrots -Spinach - Radishes- Green Onions - Beets - Chard
	April - June	Apricots - Cherries - Nectarines - Limes - Loquats - Honeydew - Watercress	Artichokes - Lettuce - Peas - Green Beans - Arugula - Spring Onions
Summer	June - August	Blueberries - Peaches- Watermelon -Plums- Cantaloupe - Fig - Jackfruit	Bell Peppers - Corn - Cucumbers -Tomatoes- Swiss Chard - Kale - Kohlrabi
	July - September	Blackberries - Figs - Raspberries - Melons - Papayas - Peach - Nectarine	Zucchini - Eggplant - Garlic - Okra - Shallots - Celery - Endive
Fall	September - November	Cranberries - Pears - Apples - Grapes - Persimmons - Quince - Dragon Fruit	Broccoli - Cauliflower - Pumpkins - Squash - Mushrooms -Turnips - Rutabaga
	October - December	Grapefruit - Kiwifruit - Lemons - Pomegranates - Quince - Satsuma - Tangelo	Sweet Potatoes - Brussels Sprouts - Parsnips - Fennel - Collard Greens - Escarole
Winter	December - February	Oranges - Kiwifruit - Dates - Tangerine - Blood Oranges - Clementines - Star Fruit	Kale - Leeks - Parsnips - Turnips - Winter Squash - Chicory - Radicchio
	January - March	Clementines - Grapefruit - Papayas - Bananas - Passion Fruit - Pear - Apple	Broccoli - Cabbage - Carrots - Celery - Beet Greens - Mustard Greens - Onion

BOOK 9:
FIRE AND COOKING

Chapter 1:

Fire Creation Techniques

Embarking on the journey of mastering fire creation techniques is a fundamental skill that connects us to our ancient ancestors and enables us to harness the transformative power of fire for warmth, cooking, and survival.

Primitive Fire-Starting Techniques

Begin with an exploration of primitive methods used by our ancestors. Learn the art of friction fire starting, where rubbing two pieces of wood together generates enough heat to ignite tinder. Discover variations such as the hand drill and bow drill, understanding the nuances of materials and techniques.

Step-by-Step Guide:

1. Select appropriate materials: a spindle, fireboard, and tinder.

2. Position the spindle on the fireboard and create downward pressure while rotating it with the hands or using a bow.

3. Generate friction by rapidly moving the spindle, creating heat.

4. Collect the ember produced and carefully transfer it to a tinder bundle.

5. Gently blow on the ember, coaxing it into a flame.

Flint and Steel

Journey into the era of flint and steel, a classic fire-starting technique. Understand how striking flint against steel produces sparks, igniting char cloth or other suitable tinder.

Step-by-Step Guide:

1. Gather materials: flint, steel, and char cloth.

2. Hold the steel firmly against the flint.

3. Strike the steel against the flint at a sharp angle to produce sparks.

4. Direct sparks onto the char cloth, which will begin to smolder.

5. Transfer the smoldering char cloth to a prepared tinder bundle and gently blow to create a flame.

Fire Pistons

Explore the concept of fire pistons, devices that use rapid compression of air to generate heat and ignite tinder. Understand the components and techniques involved in utilizing a fire piston effectively.

Step-by-Step Guide:

1. Assemble the fire piston device.

2. Place a small piece of tinder in the piston's recess.

3. Rapidly push down on the piston, compressing air inside.

4. The rapid compression creates heat, igniting the tinder.

5. Transfer the burning tinder to a larger tinder bundle for sustained flame.

Magnifying Glass or Lens

Harness the power of the sun with magnifying glass or lens fire starting. Learn how to focus sunlight onto tinder to initiate combustion.

Step-by-Step Guide:

1. Collect dry, fine tinder material.

2. Hold the magnifying glass or lens at an angle to concentrate sunlight onto the tinder.

3. Adjust the angle until a pinpoint of light forms on the tinder.

4. Once the tinder begins to smolder, transfer it to a prepared tinder bundle.

5. Blow gently to encourage the smoldering tinder to ignite.

Fire Starters and Ferrocerium Rods

Explore modern fire-starting tools such as ferrocerium rods. Understand the science behind these tools and learn to use them with various tinder materials.

Step-by-Step Guide:

1. Gather tinder material, ensuring it is dry and finely processed.

2. Hold the ferrocerium rod in one hand and the striker in the other.

3. Position the rod close to the tinder and scrape the striker down its length.

4. Sparks generated should land on the tinder, initiating combustion.

5. Transfer the burning tinder to a prepared tinder bundle for sustained flame.

Battery and Steel Wool

Delve into an unconventional but effective fire-starting method using a battery and steel wool. Understand the chemical reaction that occurs, creating heat and igniting the steel wool.

Step-by-Step Guide:

1. Obtain a small piece of steel wool and a battery (9-volt works well).

2. Stretch out the steel wool to expose more surface area.

3. Touch the battery terminals to the steel wool, completing the circuit.

4. The electrical current passing through the steel wool generates heat, causing it to ignite.

5. Transfer the burning steel wool to a prepared tinder bundle.

Fueling and Maintaining the Fire

Once the fire is ignited, understand the importance of adding appropriate fuel to sustain and build the flame. Explore different types of fire lays, such as teepee, log cabin, or pyramid, each serving specific purposes in maintaining a controlled and lasting fire.

Key Considerations:

- Gradually add larger fuel materials to the initial flame.
- Arrange the fuel in a way that allows for adequate airflow.
- Be mindful of the type of wood being used; dry, seasoned wood burns more efficiently.
- Monitor the fire and adjust fuel as needed to maintain a desired size and intensity.

Chapter 2:

Off-Grid Cooking Solutions

Embarking on the journey of off-grid cooking is a liberating venture that allows individuals to disconnect from traditional kitchen setups and explore self-sufficiency in preparing meals. Designing and building off-grid cooking solutions is not just about functionality but also about embracing a lifestyle that prioritizes independence and sustainability.

Outdoor Rocket Stove

An outdoor rocket stove is a highly efficient, wood-burning cooking solution. It uses minimal wood fuel while maximizing heat output for cooking.

Step-by-Step Guide:

1. Gather materials: bricks, metal grate, and a steel pipe.

2. Arrange the bricks to create a vertical chamber with an opening at the bottom.

3. Place the metal grate over the opening to support the cooking pot.

4. Insert the steel pipe into the side of the brick structure as a chimney.

5. Start a small fire at the bottom, feeding wood through the opening to maintain the flame.

Solar Cooker

Harnessing the power of the sun, a solar cooker uses reflective surfaces to concentrate sunlight for cooking without any fuel.

Step-by-Step Guide:

1. Create a reflective surface using materials like aluminum foil or reflective sheets.

2. Shape the reflective material into a parabolic or box-like structure.

3. Place the cooking vessel at the focal point where sunlight is concentrated.

4. Adjust the angle of the cooker throughout the day to maximize sunlight exposure.

5. Patience is key; solar cookers work best on sunny days with clear skies.

DIY Portable Camping Stove

A portable camping stove is a versatile off-grid cooking solution for outdoor enthusiasts. It's lightweight, compact, and can be fueled by various sources.

Step-by-Step Guide:

1. Gather materials: a metal container, small metal grate, and heat-resistant material.

2. Cut an opening in the container for the cooking pot.

3. Place the metal grate inside the container to support the pot.

4. Fill the bottom with a heat-resistant material like sand.

5. Use fuel tablets, twigs, or small branches for cooking.

Outdoor Clay Oven (Cob Oven)

An outdoor clay oven, also known as a cob oven, is a traditional yet effective way to bake and cook using natural materials.

Step-by-Step Guide:

1. Create a foundation using bricks or stones.

2. Mix clay, sand, and straw to form a cob mixture.

3. Build the oven structure with the cob mixture, leaving an opening for the entrance.

4. Allow the cob to dry and harden.

5. Once dry, light a small fire inside to further cure the oven before baking.

DIY Campfire Grill

A simple and versatile solution, a campfire grill allows for open-flame cooking over a controlled fire.

Step-by-Step Guide:

1. Find two sturdy logs or stones to serve as the base.

2. Place a metal grate or grill rack over the logs or stones.

3. Adjust the height of the grate by propping it up with additional stones.

4. Build a fire underneath the grate and use it for grilling or cooking in pots and pans.

Propane Camping Stove

A propane camping stove is a portable and convenient off-grid cooking solution that runs on propane canisters.

Step-by-Step Guide:

1. Choose a propane camping stove with a stable base.

2. Attach a propane canister to the stove.

3. Turn the valve to release propane and use a spark igniter or lighter to ignite the burner.

4. Adjust the flame intensity using the stove's controls.

5. Use this setup in well-ventilated areas for safety.

Haybox Cooker

A haybox cooker is an energy-efficient, off-grid solution that utilizes retained heat for slow cooking.

Step-by-Step Guide:

1. Bring a pot of food to boiling on a conventional stove.

2. Place the hot pot into an insulated container filled with hay, blankets, or other insulating materials.

3. Close the lid tightly, allowing the food to continue cooking with retained heat.

4. This method is excellent for slow-cooking stews, soups, and grains.

DIY Wood-Fired Pizza Oven

A wood-fired pizza oven brings a touch of gourmet cooking to off-grid environments, providing an opportunity to bake pizzas and more.

Step-by-Step Guide:

1. Build a base using bricks or concrete blocks.

2. Create an arched dome structure using refractory bricks or clay.

3. Build a chimney to allow smoke to escape.

4. Add a door to the front of the oven for placing and retrieving food.

5. Use dry wood for the fire, achieving high temperatures for pizza baking.

Chapter 3:

Survival Recipes

Foraged Vegetable Stir-Fry

Preparation time: 15 minutes

Cooking time: 10 minutes

Servings: 2

Ingredients:

- 2 cups foraged greens (dandelion leaves, chickweed, etc.)
- 1 cup wild mushrooms
- 1 wild onion, chopped
- 2 tablespoons wild garlic, minced
- Salt and pepper as required

Directions:

1. Wash and chop foraged greens and mushrooms.

2. Heat a pan over an open flame.

3. Include wild mushrooms, wild onion, and wild garlic to the pan.

4. Stir-fry until vegetables are tender.

5. Season with salt and pepper as required.

Campfire Grilled Fish

Preparation time: 10 minutes

Cooking time: 15 minutes

Servings: 2

Ingredients:

- 2 fresh-caught fish (trout or bass)
- Wild herbs (rosemary, thyme)
- Lemon slices
- Salt and pepper

Directions:

1. Clean and gut the fresh-caught fish.

2. Season the fish with salt, pepper, and wild herbs.

3. Skewer the fish and place them over a campfire.

4. Grill until the fish is cooked through.

5. Serve with lemon slices.

Nettle and Acorn Soup

Preparation time: 15 minutes

Cooking time: 30 minutes

Servings: 2

Ingredients:

- 2 cups nettle leaves (young, tender)
- 1/2 cup acorn flour
- Wild onion, chopped
- 4 cups water
- Salt and pepper as required

Directions:

1. Boil nettle leaves in water until tender.

2. Include acorn flour and chopped wild onion.

3. Simmer until the soup thickens.

4. Season with salt and pepper.

1. Garnish with wild cheese.

Wild Tea Infusion

Preparation time: 5 minutes

Cooking time: 10 minutes

Servings: 2

Ingredients:

- Pine needles
- Wild mint leaves
- Wild berries
- Water

Directions:

1. Boil water in a pot.

2. Include pine needles, wild mint leaves, and wild berries.

3. Let it simmer for 5-10 minutes.

4. Strain and serve.

Sautéed Cattail Shoots

Preparation time: 10 minutes

Cooking time: 10 minutes

Servings: 2

Ingredients:

- Fresh cattail shoots
- Wild garlic, minced
- Olive oil
- Salt and pepper

Directions:

1. Peel and chop fresh cattail shoots.

2. Sauté shoots with minced wild garlic in a pan.

3. Drizzle with olive oil if available.

4. Season with salt and pepper.

Roasted Root Vegetables

Preparation time: 15 minutes

Cooking time: 25 minutes

Servings: 2

Ingredients:

- Foraged root vegetables (carrots, parsnips)
- Wild herbs (thyme, oregano)
- Olive oil
- Salt and pepper

Directions:

1. Peel and chop foraged root vegetables.

2. Toss with wild herbs and olive oil.

3. Roast in a makeshift oven until vegetables are tender.

4. Season with salt and pepper.

Hickory Nut Trail Mix

Preparation time: 10 minutes

Cooking time: N/A

Servings: 2

Ingredients:

- Hickory nuts
- Dried berries
- Foraged seeds (sunflower, pumpkin)
- Wild honey (optional)

Directions:

1. Combine hickory nuts, dried berries, and foraged seeds.

2. Drizzle with wild honey for sweetness.

3. Mix well and enjoy as a trail snack.

Wildflower Salad

Preparation time: 15 minutes

Cooking time: N/A

Servings: 2

Ingredients:

- Foraged edible flowers (dandelion, violets)
- Wild greens (chickweed, lamb's quarters)
- Wild vinaigrette (wild mustard, vinegar)
- Wild nuts or seeds
- Salt and pepper

Directions:

1. Rinse and pat dry foraged edible flowers and greens.

2. Toss together with wild nuts or seeds.

3. Drizzle with wild vinaigrette.

4. Season with salt and pepper.

Wild Mushroom Risotto

Preparation time: 20 minutes

Cooking time: 30 minutes

Servings: 2

Ingredients:

- 1 cup foraged wild mushrooms (chanterelles, morels)
- 1 cup foraged rice
- Wild onion, chopped

- Wild garlic, minced
- Vegetable broth
- Olive oil
- Salt and pepper as required

Directions:

1. Sauté wild mushrooms, wild onion, and wild garlic in olive oil.

2. Include foraged rice and stir to coat with the oil.

3. Gradually add vegetable broth while stirring until rice is cooked.

4. Season with salt and pepper.

Acorn Pancakes

Preparation time: 15 minutes

Cooking time: 15 minutes

Servings: 2

Ingredients:

- 1 cup acorn flour
- Wild berries (for topping)
- Water
- Wild honey (optional)
- Wild nuts (chopped, optional)

Directions:

1. Mix acorn flour with water to make a batter.

2. Cook pancakes on a hot surface until golden brown.

3. Top with wild berries, wild honey, and chopped nuts if available.

Cattail Pollen Pancakes

Preparation time: 15 minutes

Cooking time: 15 minutes

Servings: 2

Ingredients:

- Cattail pollen
- Foraged flour
- Water
- Wild berries (for topping)
- Wild honey (optional)

Directions:

1. Mix cattail pollen with foraged flour and water to make a batter.

2. Cook pancakes on a hot surface until golden brown.

3. Top with wild berries and drizzle with wild honey if desired.

Dried Fruit Leather

Preparation time: 15 minutes

Cooking time: 6-8 hours (drying time)

Servings: 2

Ingredients:

- Assorted wild fruits (berries, apples, etc.)
- Wild honey (optional)

Directions:

1. Puree wild fruits and sweeten with wild honey if needed.

2. Spread the mixture thinly on a flat surface to dry.

3. Allow to air-dry or use sunlight until it becomes a fruit leather.

Wild Mint Tea and Biscuits

Preparation time: 15 minutes

Cooking time: 15 minutes

Servings: 2

Ingredients:

- Wild mint leaves
- Foraged flour
- Wild honey (optional)
- Water
- Wild nuts (optional)

Directions:

1. Brew tea using fresh wild mint leaves.

2. Prepare biscuits using foraged flour and water.

3. Sweeten tea with wild honey and enjoy with biscuits.

4. Garnish with wild nuts if available.

Rosehip and Pine Needle Infused Tea

Preparation time: 15 minutes

Cooking time: 10 minutes

Servings: 2

Ingredients:

- Rosehips
- Pine needles
- Water
- Wild honey (optional)

Directions:

1. Collect rosehips and pine needles.

2. Boil water and steep rosehips and pine needles to make a fragrant tea.

3. Sweeten with wild honey if desired.

Sassafras Root Beer

Preparation time: 30 minutes

Cooking time: 15 minutes

Servings: 2

Ingredients:

- Sassafras roots
- Wintergreen leaves (optional)
- Birch bark (optional)
- Water
- Wild honey (optional)

Directions:

1. Collect sassafras roots, wintergreen leaves, and birch bark.

2. Boil the roots, leaves, and bark in water.

3. Allow it to cool, strain, and sweeten with wild honey if desired.

Wild Edible Flower Fritters

Preparation time: 20 minutes

Cooking time: 15 minutes

Servings: 2

Ingredients:

- Foraged edible flowers (dandelion, pansy)
- Foraged flour
- Water
- Wild honey (optional)
- Wild nuts (chopped, optional)

Directions:

1. Mix foraged flour with water to make a batter.

2. Dip edible flowers in the batter and fry until golden brown.

3. Drizzle with wild honey and sprinkle with chopped wild nuts if desired.

Spruce Tip Syrup

Preparation time: 10 minutes

Cooking time: 15 minutes

Servings: 2

Ingredients:

- Spruce tips
- Water
- Wild honey
- Lemon juice (optional)

Directions:

1. Collect fresh spruce tips.

2. Boil spruce tips in water until a syrupy consistency is achieved.

3. Sweeten with wild honey and add lemon juice for extra flavor if desired.

Amaranth and Purslane Salad

Preparation time: 15 minutes

Cooking time: N/A

Servings: 2

Ingredients:

- Foraged amaranth leaves
- Foraged purslane
- Wild vinaigrette (wild mustard, vinegar)
- Wild nuts or seeds
- Salt and pepper

Directions:

1. Rinse and pat dry foraged amaranth leaves and purslane.

2. Toss together with wild vinaigrette.

3. Garnish with wild nuts or seeds.

4. Season with salt and pepper.

Hazelnut and Maple Energy Bars

Preparation time: 20 minutes

Cooking time: 15 minutes

Servings: 2

Ingredients:

- Hazelnuts, finely chopped
- Foraged oats
- Wild honey
- Maple syrup
- Dried berries (optional)

Directions:

1. Mix finely chopped hazelnuts, foraged oats, wild honey, and maple syrup.

2. Form into bars and let them set.

3. Include dried berries for extra flavor if available.

Birch Sap Lemon Sorbet

Preparation time: 15 minutes

Cooking time: N/A

Servings: 2

Ingredients:

- Birch sap
- Freshly squeezed lemon juice
- Wild honey (optional)
- Mint leaves for garnish

Directions:

1. Combine birch sap, freshly squeezed lemon juice, and wild honey.

2. Freeze the mixture until it reaches a sorbet consistency.

3. Garnish with fresh mint leaves before serving.

BOOK 10: NAVIGATION AND COMMUNICATION

Chapter 1:

Advanced Navigation Skills

Navigating through diverse landscapes demands a nuanced skill set beyond the basics. Advanced navigation skills are the key to confidently exploring different settings, ensuring you can find your way in any environment. This compilation of techniques empowers you to navigate with precision, whether you're in a dense forest, open desert, or mountainous terrain.

1. **Celestial Navigation:** Look to the skies for guidance. Understanding celestial bodies such as the sun, moon, and stars can serve as reliable reference points. Learn how to identify constellations and use them to determine direction, especially in open areas with unobstructed views of the sky.

2. **Shadow Stick Method:** Harness the power of the sun's movement. Drive a stick into the ground, and mark the tip of its shadow. After a set interval, mark the new shadow tip. The line connecting these points east-west, aiding your orientation based on the sun's path.

3. **Handrail Navigation:** In dense forests or areas with prominent features, use handrails like rivers, ridges, or trails as guides. By keeping these features within sight, you establish a reliable path to follow, reducing the risk of getting disoriented in complex landscapes.

4. **Dead Reckoning:** When in motion, dead reckoning estimates your current position based on a previously known location. Combine knowledge of your starting point, distance traveled, and directional changes to approximate your position, especially useful in featureless terrains.

5. **Terrain Association:** Study the topography around you. Identify prominent features, such as hills, valleys, and distinctive landmarks. Correlate these with your map to create a mental image of your location, enhancing your situational awareness.

6. **Star Navigation:** During clear nights, stars become celestial landmarks. Polaris, the North Star, is a fixed point in the northern hemisphere. Learn to identify it and use it for direction, combining this skill with knowledge of other constellations for added precision.

7. **Natural Navigation:** Master the art of observing nature's cues. Anticipate changes in vegetation, animal behavior, or patterns in rock formations to gauge your location. The environment itself can provide subtle hints about cardinal directions and potential water sources.

8. **GPS Waypoint Navigation:** Incorporate technology into your toolkit. Utilize GPS devices to set waypoints, marking specific locations of interest. This method is particularly effective when navigating vast terrains, providing accurate coordinates for precise travel.

9. **Cross-Bearing Navigation:** Combine bearings from multiple known points to pinpoint your location. By identifying landmarks on your map and aligning them with your compass, you create intersecting lines that narrow down your position, enhancing accuracy.

10. **Sun Compass:** Employ the sun as a reliable compass during daylight hours. By marking the initial and subsequent positions of your shadow at intervals, you can determine east and west, aiding your orientation without the need for additional tools.

11. **Astronomical Navigation:** For those well-versed in celestial observation, celestial bodies like the planets and moon can offer directional clues. Understanding their patterns and movements enhances your ability to navigate using the celestial sphere.

12. **Micro-Navigation:** Zoom in on the details. Micro-navigation involves navigating in small increments, paying attention to minute features like rocks, trees, or bends in a trail. This meticulous approach ensures you stay on course even in intricate environments.

13. **Sound Navigation:** Listen to your surroundings. Rivers, wind patterns, and even distant traffic can serve as auditory cues. By recognizing and interpreting these sounds, you can gain insights into your surroundings and navigate with enhanced awareness.

14. **Snow Navigation:** In snowy landscapes, traditional features may be obscured. Learn techniques specific to snowy environments, such as using snow drift patterns, tree shapes, or identifying subtle variations in terrain to maintain accurate navigation.

15. **GPS Track Navigation:** Modern GPS devices allow you to record and follow pre-existing tracks. Learn how to download or create tracks for your routes, ensuring you can retrace your steps accurately or follow established paths in unfamiliar terrain.

16. **Time and Distance Estimation:** Develop a sense of time and distance traveled. By maintaining a steady pace and estimating the time it takes to cover specific distances, you can approximate your position, especially when traditional landmarks are scarce.

17. **Polar Navigation:** In polar regions, where magnetic compasses may be unreliable, use alternative methods like the sun compass or star navigation. Understanding the unique challenges of polar landscapes is crucial for precise navigation.

18. **GPS Mapping Applications:** Leverage the capabilities of GPS mapping apps on smartphones. These apps often provide detailed topographic maps, waypoints, and tracking features, enhancing your navigation capabilities in a convenient, portable device.

19. **Mental Mapping:** Sharpen your mental map. Continually visualize your route, landmarks, and key features. Cultivating a mental map enables you to navigate confidently even when physical maps or devices are unavailable.

20. **Sun and Moon Navigation:** Beyond their role in star navigation, the sun and moon offer additional guidance. Learn to interpret the sun's path throughout the day and the moon's phases, enhancing your ability to determine direction and time.

Chapter 2:

Emergency Signal Crafting

When venturing into the wild, being prepared for unexpected situations is paramount. Among the essential skills for survival is the ability to craft and use emergency signals effectively. These signals serve as a lifeline, communicating your distress or location to potential rescuers.

Visual Signals

Visual signals are a universal language in the wilderness, catching the attention of potential rescuers from a distance. Master the art of crafting visible signs that can be spotted in various terrains.

- **Bright Clothing or Objects:** Wear or display bright-colored clothing, especially in contrast to the surroundings. Attach reflective materials to increase visibility, making yourself easily noticeable.
- **Signal Fires:** Create a signal fire using available resources. Use green vegetation to produce smoke, adding a distinctive visual element to the fire. Three equally spaced smoke signals signify distress.
- **Ground-to-Air Signals:** Arrange objects or create patterns on the ground that can be easily seen from the air. Utilize contrasting materials like rocks or branches to form clear shapes, such as an "X" or a large arrow.
- **Mirror Signals:** Carry a small signaling mirror in your kit. Flash sunlight toward aircraft or distant observers to create reflective signals. Aim the mirror's reflection at your intended target.

Auditory Signals

Sound can travel significant distances in the wilderness. Crafting effective auditory signals increases your chances of being heard by potential rescuers.

- **Whistles:** Carry a whistle with you at all times. Three short bursts followed by a pause is a recognized distress signal. Use different patterns, such as long blasts or series of short and long blasts, to convey specific messages.

- **Shouts and Calls:** Vocalize loudly and clearly. Use distinct patterns, like shouting three times in succession, to indicate distress. Repeating the signal at regular intervals is crucial for catching attention.
- **Use of Noise-Making Objects:** If available, employ noise-making objects like metal containers to create rhythmic sounds. This can be particularly effective in areas with echoes, maximizing the reach of your auditory signal.
- **Fire Crackers or Whistling Arrows:** If you have access to items that produce loud sounds, such as firecrackers or arrows with attached whistling devices, use them judiciously to attract attention.

Communication Through Objects

Crafting signals using physical objects can convey information about your location, condition, or the urgency of your situation.

- **Ground Markings:** Arrange rocks, logs, or other materials to create visible patterns or write distress messages on the ground. Make your signals large and clear enough to be spotted from a distance.
- **Signal Panels:** If you have a brightly colored item like a jacket or tarp, use it as a signal panel. Wave it to attract attention or affix it to an elevated position for increased visibility.
- **Use of Personal Items:** Strategically place personal items, such as backpacks or equipment, to create noticeable signals. Ensure they stand out against the natural surroundings.
- **Creating Symbols or Codes:** Develop simple symbols or codes that can convey specific messages. For example, arranging rocks in a triangle may indicate your need for medical assistance.

Technology-Assisted Signals

Leverage available technology to enhance your signaling capabilities. While technology may not always be reliable, incorporating it into your signaling strategy can provide an extra layer of communication.

- **Flashlights or Headlamps:** In low-light conditions, use flashes of light from flashlights or headlamps to signal. Employ Morse code if possible, spelling out SOS or other recognizable distress signals.
- **Cell Phones or Radios:** If you have functional communication devices, use them to call for help. Even if you don't have a signal for a voice call, attempt to send text messages or use emergency beacon features if available.

- **GPS Devices:** If your GPS device has tracking features, activate them to share your real-time location with emergency services or contacts. Familiarize yourself with the capabilities of your specific device.

Environmental Adaptation

Crafting signals that blend seamlessly with the environment requires resourcefulness. Adapt your signaling methods to suit the specific conditions of the wilderness.

- **Snow Signals:** In snowy environments, use dark-colored materials or create patterns that contrast with the snow. Utilize ski poles, branches, or even dark clothing to form visible signals.
- **Sand or Desert Signals:** In sandy or desert settings, arrange rocks or use materials that contrast with the light-colored sand. High-contrast signals stand out against the desert landscape.
- **Forest Signals:** Employ natural materials like leaves, branches, or even crushed vegetation to create visible signals in forested areas. Use materials that contrast with the predominant colors of the surroundings.

Persistence and Consistency

Consistency is crucial when crafting emergency signals. Repeat your signals at regular intervals and maintain persistence, as rescue efforts may take time.

- **Establishing Patterns:** Set a pattern for your signals, whether visual or auditory. This helps rescuers recognize intentional signals from natural occurrences.
- **Day and Night Signals:** Continue signaling during both day and night. Adjust your techniques to account for changes in visibility, emphasizing lights, fires, or reflective signals during the night.
- **Signaling Across Days:** If your situation extends over several days, devise a signaling routine. Consistent signaling increases the likelihood of being noticed by search and rescue teams.

Chapter 3:

Technology in Survival

In today's world, technology is deeply woven into every aspect of our lives. It's instinctive to consider how these progressions can be utilized for survival in the expansive wilderness. Navigating the natural environment becomes safer with the thoughtful application of technology, especially Global Positioning System (GPS) devices and radios, turning a potential danger into a skillfully managed escapade.

GPS Technology in the Wilderness

GPS technology has revolutionized navigation, providing accurate positioning information anywhere on the planet. A typical GPS device consists of satellites orbiting Earth and a receiver on the ground. By triangulating signals from multiple satellites, the device calculates your precise location.

Applications in Survival:

1. **Mapping and Navigation:** GPS devices offer detailed topographic maps, allowing you to plan routes, mark waypoints, and track your movements. This is invaluable in unfamiliar terrain or when exploring off the beaten path.

2. **Emergency Beacons:** Many modern GPS devices come equipped with emergency beacon features. Activating these beacons sends distress signals to search and rescue teams, providing them with your exact location for swift assistance.

3. **Geo-Caching and Resource Location:** Utilize GPS for geo-caching, a recreational activity that combines outdoor exploration with treasure hunting. Additionally, mark locations of valuable resources like water sources, shelter spots, or wildlife observation points.

4. **Real-Time Tracking:** Some GPS devices allow for real-time tracking, enabling loved ones or rescue teams to monitor your location remotely. This can be a reassuring feature for solo adventurers or those embarking on extended trips.

5. **Sun and Moon Data:** Advanced GPS models may provide data on sunrise, sunset, moon phases, and celestial events. This information aids in planning activities, maximizing daylight, and navigating during low-light conditions.

Radios as Lifelines in the Wilderness

Radios serve as reliable communication tools in the wilderness, facilitating contact with fellow adventurers or emergency services. Two-way radios (walkie-talkies) and emergency radios are commonly used, each with distinct advantages.

Effective Communication Strategies:

1. **Two-Way Radios:** These compact devices allow short-distance communication between users. Ensure your group has a set of tuned radios, and establish communication protocols before embarking on any journey.

2. **Emergency Radios:** Equipped with features like NOAA weather alerts and hand-crank power, emergency radios are essential for staying informed about changing weather conditions. Some models also include built-in flashlights and charging ports for other devices.

3. **Frequencies and Channels:** Familiarize yourself with radio frequencies and channels relevant to your location. In emergencies, you may need to switch to designated emergency frequencies or channels to call for assistance.

4. **Antenna Adjustments:** If using two-way radios in mountainous or dense forest areas, adjusting the antenna length can enhance signal transmission. Experiment with different lengths to find the optimal configuration.

5. **Emergency Signaling:** In dire situations, radios can be used for Morse code signaling. Learn the basic Morse code for distress signals, such as SOS, to communicate your need for help effectively.

Challenges and Considerations

Battery Life and Power Sources:

- **Battery Conservation:** Efficiently manage the battery life of your GPS and radios by turning them off when not in use. Carry spare batteries or portable chargers to extend usage during extended trips.
- **Solar Chargers:** Embrace solar-powered chargers to replenish the energy of your devices. These chargers harness the sun's energy and can be a sustainable solution in sunny conditions.

Terrain and Obstructions:

- **Radio Signal Interference:** Understand that radio signals can be affected by natural barriers like mountains or dense foliage. Choose high points or clearings for better transmission and reception.
- **GPS Signal Reliability:** While GPS technology is generally reliable, signals may be compromised in deep canyons or thick tree cover. In such scenarios, a combination of GPS and traditional navigation methods is prudent.

Weather Conditions:

- **Waterproofing:** Invest in waterproof or water-resistant cases for your GPS and radios to protect them from rain or accidental submersion. Being exposed to the elements can compromise their functionality.
- **Weather Impact on Signal Strength:** Be aware that adverse weather conditions, such as heavy rain or storms, can affect radio signals and reduce their range. Exercise caution and adapt communication strategies accordingly.

Integrating Technology into Your Survival Plan:

Pre-Trip Planning:

- **Device Familiarization:** Before embarking on any wilderness adventure, familiarize yourself with the functionalities of your GPS and radios. Practice using emergency features and navigation tools.
- **Map Preloading:** Preload relevant maps onto your GPS device. Having detailed maps of your intended route, waypoints, and nearby resources enhances your navigational capabilities, especially in areas with limited signal reception.

Communication Protocols:

- **Establish Clear Protocols:** Define clear communication protocols within your group. Establish designated channels, codes, or signals for routine check-ins, emergencies, and navigation updates.
- **Emergency Contact Information:** Program emergency contact information into your GPS and radios. In the event of an emergency, this ensures that essential details are readily available when calling for help.

Regular Maintenance:

- **Device Checks:** Regularly check the condition of your devices, including the integrity of antennas, battery levels, and overall functionality. Identify and address any issues before heading into the wilderness.
- **Software Updates:** Keep your GPS firmware and mapping software up to date. Software updates may enhance performance, fix bugs, and introduce new features that can be beneficial in survival scenarios.

BOOK 11:
TOWARDS COMPLETE
SELF-SUFFICIENCY

Chapter 1:

Transitioning to Off-Grid Living

Transitioning to off-grid living is a transformative journey toward self-reliance and sustainability. As you embark on this path, practical steps become the building blocks of your independence. This lifestyle shift requires thoughtful consideration and a commitment to reducing dependence on external resources. Here, we explore tangible steps to guide your transition to off-grid living, empowering you to embrace a life less tethered to traditional utilities and systems.

1. **Assessing Your Energy Needs**

Understanding Consumption:

- **Energy Audit:** Begin by conducting an energy audit to understand your current consumption. Identify which appliances and devices contribute most to your energy usage.
- **Prioritizing Efficiency:** Prioritize energy-efficient appliances and lighting. Transitioning to LED bulbs, energy-star rated devices, and appliances can significantly reduce your overall energy demands.
- **Investing in Renewable Energy:** Explore renewable energy sources like solar panels and wind turbines. Determine the feasibility of installing these systems based on your location and energy requirements.

2. **Water Sourcing and Management**

Sustainable Water Practices:

- **Assessing Water Needs:** Evaluate your water needs, considering daily consumption, gardening, and livestock. This assessment forms the basis for implementing sustainable water practices.
- **Rainwater Harvesting:** Harvest rainwater for household use and irrigation. Install rain barrels or tanks to collect and store rainwater, reducing reliance on conventional water sources.
- **Greywater Systems:** Implement greywater systems to recycle and reuse water from daily activities like laundry and dishwashing. This conserves water and minimizes environmental impact.

3. **Establishing Sustainable Agriculture**

Homegrown Food Production:

- **Creating a Kitchen Garden:** Start a kitchen garden to grow fresh fruits, vegetables, and herbs. This not only provides a sustainable food source but also enhances your connection to the land.
- **Composting:** Establish a composting system to repurpose kitchen and garden waste, enriching the soil and diminishing the reliance on chemical fertilizers. This fosters robust plant growth and contributes to a healthier environment.
- **Permaculture Principles:** Explore permaculture principles to design an integrated and sustainable food production system. Embrace concepts like companion planting and natural pest control.

4. **Building a Resilient Shelter**

Sustainable Shelter Practices:

- **Energy-Efficient Construction:** If building a new home, prioritize energy-efficient construction materials and designs. Consider passive solar design principles to optimize natural heating and cooling.
- **Off-Grid Power Systems:** Integrate your energy sources into the construction plan. Ensure the home is designed to accommodate solar panels or wind turbines, optimizing their efficiency.
- **Natural Building Techniques:** Explore natural building techniques using materials like straw bales, cob, or recycled materials. These methods minimize the environmental impact and contribute to a healthier living space.

5. **Waste Reduction and Recycling**

Minimizing Environmental Impact:

- **Zero-Waste Lifestyle:** Strive for a zero-waste lifestyle by minimizing single-use items and embracing reusable alternatives. Committed waste reduction reduces your ecological footprint.
- **Recycling Practices:** Establish a recycling system for materials that cannot be reused or composted. Familiarize yourself with local recycling programs and explore creative ways to repurpose items.
- **Upcycling and Repurposing:** Embrace upcycling and repurposing as part of your lifestyle. Turn discarded items into functional objects or use them for DIY projects, reducing the need for new purchases.

6. **Sustainable Transportation**

Environmentally Conscious Mobility:

- **Bicycling:** Integrate bicycling into your daily routine for short-distance commuting. This eco-friendly mode of transportation promotes personal health and reduces your carbon footprint.
- **Electric Vehicles (EVs):** If possible, consider transitioning to electric vehicles. EVs contribute to cleaner air and are a more sustainable option compared to traditional gasoline-powered cars.
- **Carpooling and Public Transport:** Explore carpooling options or utilize public transportation to further reduce your reliance on personal vehicles. Community-based solutions contribute to sustainable mobility.

7. **Off-Grid Communication and Connectivity**

Embracing Digital Independence:

- **Alternative Internet Solutions:** Investigate alternative internet solutions, such as satellite internet or mobile hotspots, to maintain connectivity without relying on traditional providers.
- **Off-Grid Communication Devices:** Consider off-grid communication devices like satellite phones or radios for emergency situations. These devices provide communication options beyond conventional networks.
- **Digital Detox Practices:** Implement digital detox practices to balance connectivity and independence. Designate specific times for unplugging and connecting with nature or engaging in analog activities.

Chapter 2:

Renewable Energy Projects

Starting renewable energy projects allows individuals to assert control over their energy consumption, minimize environmental impact, and adopt sustainable living practices. This resource furnishes detailed do-it-yourself instructions for constructing solar, wind, and hydro systems. Engaging in these initiatives not only supports a more eco-friendly future but also provides a practical learning opportunity, enabling the generation of clean energy with efficiency.

Solar Power System

Step-by-Step Guide:

1. **Solar Panel Selection:** Begin by selecting high-quality solar panels that suit your energy needs. Consider factors such as efficiency, durability, and available space for installation.

2. **Calculating Energy Needs:** Conduct an energy audit to determine your daily energy consumption. This calculation serves as the foundation for determining the number of solar panels required.

3. **Mounting Solar Panels:** Install solar panels in a location that receives maximum sunlight exposure. Optimal angles and orientations enhance energy capture. Roof mounts or ground mounts are common installation options.

4. **Wiring and Inverter Installation:** Link the solar panels either in a series or parallel configuration, based on your system's design. Integrate an inverter to transform the direct current (DC) generated by the panels into practical alternating current (AC) to power your household.

5. **Battery Bank Setup:** If opting for off-grid systems, incorporate a battery bank to store excess energy generated during sunny periods. Deep-cycle batteries are suitable for this purpose.

6. **Charge Controller Installation:** Include a charge controller to regulate the charging of batteries. This prevents overcharging, extending the life of your battery bank.

7. **Monitoring System Performance:** Implement a monitoring system to track energy production and consumption. Online platforms or specialized monitoring devices provide real-time data.

Wind Turbine System

Step-by-Step Guide:

1. **Wind Resource Assessment:** Evaluate the wind resource at your location to determine the feasibility of a wind turbine system. Online wind maps and anemometer readings aid in this assessment.

2. **Selecting a Wind Turbine:** Choose a wind turbine model based on your wind resource assessment. Horizontal-axis and vertical-axis turbines are common types, each with unique advantages.

3. **Tower Installation:** Install a sturdy tower at a height that maximizes wind exposure. Taller towers capture stronger and more consistent winds. Establish a secure and steady base for the tower.

4. **Wiring and Charge Controller:** Connect the wind turbine to a charge controller. This device regulates the charging of batteries, preventing overcharging and ensuring efficient energy storage.

5. **Battery Bank and Inverter:** Similar to solar systems, incorporate a battery bank for energy storage. Install an inverter to convert the generated DC into usable AC for household appliances.

6. **Safety Measures:** Implement safety features such as a braking system to protect the turbine during high winds. Consistent routine inspections guarantee peak efficiency and extended lifespan.

7. **Monitoring and Adjustments:** Set up a monitoring system to track the performance of the wind turbine. Regularly assess and adjust the system to maximize efficiency.

Hydroelectric Power System

Step-by-Step Guide:

1. **Water Resource Assessment:** Assess the water flow and elevation drop in a nearby stream or river. A higher flow and elevation drop result in more energy potential.

2. **Choosing a Turbine Type:** Select a turbine type based on your water resource assessment. Common types include impulse turbines (for high head) and reaction turbines (for low head).

3. **Diversion Channel Construction:** If using a low head system, create a diversion channel to direct water flow through the turbine. Ensure proper channel design for efficient energy capture.

4. **Turbine Installation:** Install the chosen turbine in the diversion channel or directly in the flowing water. Position the turbine to capture the maximum energy from the water flow.

5. **Generator and Wiring:** Connect the turbine to a generator using appropriate wiring. The generator transforms the mechanical energy from the turbine into electrical energy.

6. **Battery Bank and Inverter:** Include a battery bank for storing excess energy generated by the hydro system. Set up an inverter to transform direct current (DC) into alternating current (AC) for domestic applications.

7. **Safety Measures:** Implement safety measures such as screens to prevent debris from entering the turbine. Routine examinations and upkeep are essential for ensuring prolonged functionality.

8. **Monitoring and Optimization:** Set up a monitoring system to track the hydro system's performance. Regularly optimize the system based on seasonal variations in water flow.

Chapter 3:

Designing a Sustainable Homestead

Embarking on the path of creating a sustainable homestead transforms one's lifestyle by incorporating self-sufficiency elements. This approach requires thoughtful decision-making, resourcefulness, and fostering a balanced connection with the environment. Here, we delve into crucial aspects of building a sustainable homestead, offering a plan for those aiming for a more resilient and eco-friendly life.

1. **Energy Independence**

Harnessing Renewable Sources:

- **Solar Power Integration:** Embrace solar panels to harness the abundant energy from the sun. Strategically position solar arrays to maximize exposure and install energy-efficient appliances to optimize consumption.
- **Wind Turbines:** Explore the possibility of installing wind turbines on your homestead. These turbines convert wind energy into electricity, providing an additional renewable energy source.
- **Hydroelectric Systems:** If applicable, consider harnessing the power of flowing water on your property through small-scale hydroelectric systems. These systems can contribute to your energy needs, especially in areas with a water source.

2. **Sustainable Water Management**

Preserving and Harvesting Water:

- **Rainwater Harvesting:** Set up systems to collect and store rainwater, promoting sustainability and decreasing dependence on external water sources for diverse purposes.
- **Greywater Recycling:** Implement greywater systems to recycle and reuse water from daily activities like bathing and laundry. This minimizes water wastage and enhances water efficiency.
- **Water Conservation Practices:** Embrace water-efficient appliances, adopt practices like drip irrigation in your garden, and fix leaks promptly to ensure responsible water use.

3. **Organic Food Production**

Cultivating a Bountiful Garden:

- **Permaculture Principles:** Design your garden based on permaculture principles that emphasize sustainable, regenerative practices. Integrate companion planting, mulching, and natural pest control techniques.
- **Composting:** Establish a composting system to recycle organic waste from your kitchen and garden. Compost enriches the soil, promoting healthy plant growth without the need for synthetic fertilizers.
- **Diverse Crop Selection:** Plant a variety of crops to promote biodiversity and resilience. Consider heirloom and native plant varieties adapted to your specific climate and soil conditions.

4. **Eco-Friendly Shelter**

Sustainable Construction Methods:

- **Natural Building Materials:** Examine the utilization of eco-friendly and regionally obtained construction materials like straw bales, adobe, or repurposed wood. These alternatives present reduced environmental footprints when contrasted with traditional choices.
- **Energy-Efficient Design:** Implement passive solar design principles to optimize natural heating and cooling. Proper insulation, orientation, and window placement contribute to energy efficiency.
- **Off-Grid Options:** Consider integrating off-grid systems such as composting toilets, rainwater harvesting, and renewable energy to reduce dependence on centralized utilities.

5. **Waste Reduction and Recycling**

Minimizing Environmental Impact:

- **Zero-Waste Practices:** Strive for a zero-waste lifestyle by reducing single-use items, recycling materials, and repurposing items whenever possible.
- **Upcycling and Repurposing:** Embrace creativity by upcycling and repurposing materials. Turn discarded items into functional objects, reducing the need for new purchases.
- **Comprehensive Recycling:** Establish a comprehensive recycling system for materials that cannot be repurposed or reused. Familiarize yourself with local recycling programs and regulations.

6. **Sustainable Transportation**

Reducing Carbon Footprint:

- **Bicycle-Friendly Design:** Plan your homestead layout to facilitate bicycle commuting. Create dedicated paths and storage areas to encourage environmentally friendly transportation.
- **Electric Vehicles:** Explore the use of electric vehicles (EVs) for on-site transportation. Charging stations powered by your renewable energy sources can further reduce your carbon footprint.
- **Carpooling Initiatives:** Foster a sense of community among homesteaders to encourage carpooling. Shared transportation reduces individual energy consumption and promotes collaboration.

7. **Conservation of Natural Habitats**

Preserving Biodiversity:

- **Wildlife-Friendly Landscaping:** Design your homestead landscape to provide habitats for local wildlife. Incorporate native plants then avoid the use of harmful pesticides.
- **Natural Pest Control:** Implement natural pest control methods such as companion planting, beneficial insect attraction, and other eco-friendly practices to maintain a balanced ecosystem.
- **Erosion Prevention:** Utilize sustainable land management practices to prevent soil erosion. Plant cover crops, implement contour plowing, and design water catchment systems to retain soil fertility.

8. **Mindful Consumer Practices**

Conscious Living Choices:

- **Ethical Purchasing:** Adopt ethical and sustainable consumer practices by supporting local businesses, choosing eco-friendly products, and prioritizing fair trade options.
- **Community Engagement:** Participate in community initiatives that promote sustainable living. Collaborate with neighbors on communal projects, such as community gardens or shared resources.
- **Educational Outreach:** Share your expertise and personal experiences to benefit your community. By actively participating in educational outreach, you contribute to the development of a sustainable culture and empower individuals to make well-informed decisions.

Conclusion

Embarking on this survival journey through the diverse chapters of prepping and self-sufficiency has been enlightening. From the fundamentals of survivalism to mastering shelter construction, each section has contributed valuable insights.

In the prepping guide, understanding TEOTWAWKI scenarios broadened our perspective, preparing us for the unexpected. Bug-in versus bug-out considerations emphasized adaptability. Water, fire, and food integration emerged as essential components, further detailed in canning and food preservation techniques.

Mental health discussions underscored the importance of resilience, crucial in crises. Shelter essentials highlighted situational appropriateness, covering various scenarios from camping to winter survival.

The worst-case scenario guide tackled nuclear disasters, pandemics, natural disasters, power grid failures, and urban disruptions. The home defense guide, addressing securing the home, perimeter defense, and secret storage solutions, showcased a comprehensive approach.

The survival shelter guide offered diverse construction methods, including lean-to and snow cave shelters. DIY bunkers and anti-atomic shelters provided a deeper understanding of fortified retreats.

The self-defense and first aid crash course emphasized creating survival kits, covering essentials from shelter and warmth to navigation and communication. Mental health management became a focal point, acknowledging its significance in crisis situations.

The perfect prepper's survival checklist compiled a comprehensive list, ensuring nothing crucial is overlooked. Water mastery and sustenance and nutrition deep-dived into sourcing, purification, and foraging practices.

Fire and cooking explored primitive fire-starting techniques and off-grid cooking solutions. Survival recipes provided practical and nutritious meal ideas using foraged ingredients.

Navigation and communication delved into advanced skills and crafting emergency signals. Technology's role in survival was analyzed, highlighting its challenges and potential benefits.

Transitioning towards complete self-sufficiency involved designing a sustainable homestead and renewable energy projects, such as solar, wind, and hydroelectric power systems.

In wrapping up this self-sufficiency journey, the key lies in the meticulous application of learned skills. Embrace adaptability, prioritize mental health, and integrate sustainable practices for a self-sufficient future. This guide serves as a foundation; your journey towards self-sufficiency is an ongoing, empowering process.

Now armed with a wealth of knowledge and practical skills from this comprehensive guide to self-sufficiency, it's time to transform information into action. The call to action is clear: apply what you've learned and embrace a self-sufficient lifestyle.

First and foremost, conduct a thorough assessment of your current preparedness level. Utilize the survival checklist provided to ensure you have covered all essentials – water, food, shelter, first aid, and more. Take proactive steps to fortify your home, securing it against potential threats.

Implement the shelter construction techniques learned, whether it's a lean-to in the wilderness or fortifying your existing space. Consider the various disaster scenarios covered and tailor your preparations to your geographical location and personal circumstances.

Embrace the art of self-defense and first aid by assembling a comprehensive survival kit. Your ability to navigate, communicate, and adapt to unforeseen challenges is paramount, so practice advanced navigation skills and master emergency signal crafting.

Dive into sustainable practices and renewable energy projects to inch closer to complete self-sufficiency. Whether it's setting up a solar power system or understanding ethical hunting practices, each step brings you closer to a more resilient and independent lifestyle.

Remember, self-sufficiency is not just about surviving; it's about thriving in any situation. Embrace the knowledge gained on foraging, food preservation, and off-grid cooking to create nourishing meals from nature's bounty.

Lastly, foster a resilient mindset. Mental health management is not just a topic in the guide; it's a crucial aspect of your overall preparedness. Build resilience, adaptability, and a positive outlook to face challenges head-on.

Your journey towards self-sufficiency is a dynamic and ongoing process. Start small, but start now. Every action, no matter how minor, contributes to your preparedness. Share your knowledge with others, form a community, and collectively build a network of self-sufficient individuals.

In applying these skills, you're not just preparing for the worst; you're actively shaping a future where you have the power to overcome any challenge. The call to action is yours – seize it, embrace it, and embark on a journey towards a more self-sufficient and empowered life.

As we conclude this journey together, let's step forward with the confidence and readiness to tackle whatever lies ahead. I'm hoping this book isn't just informative, but also a source of inspiration for your off-grid journey, whether you're just starting out or you're well on your path. If something in the book strikes a chord with you, I'd be super grateful if you could take a moment to leave a honest review on Amazon. Your thoughts are incredibly precious to me, and sharing your feedback not only helps me grow but also helps others stumble upon the book and the life-enriching tips it's packed with.

And guess what? I've got some cool extras just for you – exclusive ebook and audiobook bonuses that I'm sure you'll love. Just scan the QR codes in the book to claim these goodies. I'm confident these additional resources will be a great help as you forge your path to more self-reliant and robust living. A huge thank you again for your amazing support. Enjoy your adventure!

SCAN THESE QR CODES TO GET AN EXCLUSIVE ARRAY OF BONUSES JUST FOR YOU!

DOWNLOAD YOUR EXTRAS!

Made in the USA
Columbia, SC
04 September 2024

41781629R00111